Return to Moon Cave

Robert K. Leśniakiewicz

Miloš Jesensky

ORIGINALLY PUBLISHED IN POLAND BY PUBLISHING HOUSE RADWAN
IN 2010 AS POWROT DO KSIEZYCOWEJ JASKINI

***

TRANSLATED AND PUBLISHED IN ENGLISH WITH PERMISSION.

***

PAPERBACK ISBN: 978-1-7342857-3-4
EPUB ISBN: 978-1-3936734-0-8

***

WRITTEN BY ROBERT KONSTANTY LEŚNIAKIEWICZ & DR MILOŠ JESENSKÝ
PUBLISHED BY ROYAL HAWAIIAN PRESS
COVER ART BY TYRONE ROSHANTHA
TRANSLATED BY RAFAL STACHOWSKY
PUBLISHING ASSISTANCE: DOROTA RESZKE

***

FOR MORE WORKS BY THIS AUTHOR, PLEASE VISIT:
WWW.ROYALHAWAIIANPRESS.COM

***

VERSION NUMBER 1.00

TABLE OF CONTENTS

ADMISSION

The publication of our joint book entitled the *Moon Cave Mystery* has already been accomplished. A lot has changed, but the most important change took place in the part of Europe where we live. Poland and Slovakia joined the group of countries from the Schengen Agreement, and thanks to this, the border almost completely disappeared between our countries. Our dream came true and the border really started to connect, not to divide people. This is a very good prognosis for the future, because it will be possible to undertake these works and searches that could not be carried out for administrative and bureaucratic reasons.

It is a pity that Dr. Antonin Horák did not live to see it, who started the Moon Cave Mystery with his report! For now, we gave up on exploration in the field. Simply because it requires forces and resources that we do not have, and secondly because we do not know exactly in which locations we should look for it. In principle, we have selected two:

1. Surroundings of the Šip and Kopa Mountains in the Veľká Fatra range.

2. Surroundings of Żegiestów-Żdziaryki in Poland and on the Slovak side of the border near Súlin.

Both of these locations meet basically most of the points based on the geographical description given by Dr. Horak, but not all - that is why the location of this cave is still controversial.

Personally, I am of the opinion that the matter should be resolved by us - the representatives of countries directly interested in solving this puzzle: Slovaks, Czechs, Poles and possibly Hungarians and Romanians - if the latter show interest. After all, Dr. Horak was born and raised in Transylvania in the area of the former Austro-Hungarian Empire. On the other hand, the rest of the world may eventually provide research results and data on the Moon Cave discovered and explored by us. It cannot be otherwise.

I believe that the primacy in this respect belongs to the Slovaks and Czechs because it is the territory of Slovakia, and Dr. Horak was a Czech Jew. So this should be the case.

We are currently collecting information about other strange caves in the world and the phenomena occurring in them. We have information about the discoveries made in Obłazowa Cave near Nowy Targ, where scientists have found everyday objects from about 32,000 years ago, including boomerangs that are so far the first in Poland and probably the only one on our continent. In 1985, Polish archaeologists discovered in the Obłazowa Cave, on the border of Spisz and Podhale, a concentration of valuable ceremonial objects. There was a curved, polished fragment of the mammoth's blow among them.

"I admit that I did not know at once that we were dealing with a boomerang," says Prof. Paweł Valde-Nowak from the Institute of Archeology and Ethnology of the Polish Academy of Sciences in Krakow. "At first, we thought it was a carefully made dagger. After closer research, however, we found that due to the appearance associated with the tools used by the Aborigines, it must be a boomerang, only that the Paleolithic used by man," adds Prof. Valde-Nowak.

Only in Australia, has the use of a boomerang survived to our times. So why do we not know anything about him?

"Until 1995, field work in the cave was in progress. Now in the research program, there was a break foreseen for the development of the results of the first stage and there is no decision when further examination of the sites (Cave in Obłazowa and adjacent to Obłazowa stand of population from the end of Paleolithic) will be taken," explains the scientist.

The geophysical research carried out in July in the Obłazowa cave in Pothole (Now Tar area) - one of the most important Paleolithic sites in Poland - revealed the existence of buried corridors.

In cooperation with the Institute of Geosciences of the University of Silesia, geophysical surveys have been made in the interior of the cave and its immediate surroundings. They showed the explored parts of the cave that are still collapsed or silted, extensive corridors and perhaps a large chamber. A small hollow in the field a few dozen meters from the cave turned out to be a deep karst funnel. It probably connects to the underground part of the cave. There used to be the flow of an underground river.

Obłazowa Cave was inhabited by over 50 thousand years ago and its underground parts were buried much later. Therefore, the intact traces of its oldest inhabitants could be preserved in the deeper parts. During the earlier studies of the cave, it identified 11 settlement layers. The eighth layer dated to around 30-28,000 BC was particularly interesting which contained numerous ornaments and tools made of stone and horn - including the now famous boomerang from mammoth bones. There were also truncated human fingers - the oldest human remains known from Polish lands. At that time, the cave fulfilled the role of a sanctuary.

Why is the Picture Cave so important from our point of view and what does it have in common with the Moon Cave? Well, its age is over 30,000 years and during this time, despite the earthquakes that are raging here. You must know that in the period between 1998 and 2004 AD, there were over 120 earthquakes in southern Poland, of which at least 20 had a strength of 9 degrees on the Richter scale, and in one case, this value reached M10.0. It was terrible for the inhabitants of Central Europe on 28.V.1776, when in Slovakia there was not even one brick house! And despite this powerful, destructive earthquake, the caves in Poland, Slovakia and the Czech Republic have almost remained intact!

We also took trips to two other Polish caves: Jaskinia Niedźwiedzia (800-807 m above sea level)[1] in Śnieżnik massif (1425 m) near Kłodzko in Lower Silesia and Jaskinia Raj (250-256-259 m) near Kielce. A characteristic feature of these caves is the beautiful and rich, colorful drip dress that has not been destroyed during these earthquakes. What is the conclusion? Well, it seems that there is hope that the Moon Cave did not collapse during one of these cataclysms that affected our countries, and what's more - if it exists - we can find it intact. This is a very important observation.

The second momentous observation is connected with a certain phenomenon observed in these caves and in Wąwóz Kraków - as we have already mentioned in our work - The Cracow Gorge in the Polish Western Tatras is a peculiar field formation resulting from the collapse of a cave of about 3 km long running from the Kościeliska Valley (1110 m) up to the peak of Ciemniak Mountain (2096 m). According to various sources, this cave collapsed about 10,000 - 12,000 years ago due to an unknown cataclysm.

---

[1] The heights of individual cave entrances.

On the other side of the natural section, which is the crest of Twelftylłł (1858 m), there is a gigantic Wantule boulder (from the word "wanta" or boulder). It is several hundred huge rock snowmen, which are located in the upper part of the Miętusie Valley (about 1200 - 1300 m). This boulder was created as a result of slipping on the glacial tongue of the north-eastern rock projection of Dziurawe lying in Miętusie, as a result of which it disintegrated into rock, limestone idols with a volume of a few cubic meters. And again - the age of this boulder ranges from 10 to 12 thousand years.

In both caves - Bear and Paradise - there are rocky peaks covered with infiltrations, the age of which is estimated again for 10 - 12 thousand years. So it seems that in this part of the world, there was some cataclysm which caused the collapse of caves, landslides and the formation of landmarks. This century, which repeats itself, shows a similarity to the event described by Aristocles called Plato (427-347 BC) - and of course, the destruction of the legendary Atlantis. In this case, this is one proof of more talking in relation to the relationship of Plato included in his dialogues "Timajos" and "Kritias". Hence the certainty of Dr. Horak when he spoke about Platonic civilizations in the context of the existence of the Moon Cave.

But this is not all because the matter of the existence of the Moon Cave has its second bottom, which we have already signaled in our book. It is a question of the existence of the famous underground land - Shambhalli-Agharta, which according to some authors is to be located in our planetary underground, somewhere near Central Asia or in the Tibetan area, on the border with India. This is between the Himalayas and Transhalayas, the sources of the Brahmaputra, near the giant pyramid holy mountain Kajłas / Kailas (6716 - 6666 m). [2]

---

[2] Two extreme assessments of this mountain are given here.

Known in the interwar period, Polish explorer and writer Ferdynand A. Ossendowski (1876-1945) in his bestseller, "Through the land of people, animals and gods" (Warsaw 1923), apart from the descriptions of his daring escape from cruel death from the hands of both red and white Russians, he gives a description of what he calls the "Agharty underground state". According to him, Agharta is deep underground in Tibet and Mongolia, and the masked entrances are found all over the world. It was founded 60,000 years ago and has a significant impact on what is happening in the world. If interested, I refer you to this book, which contains a description of this land. It is to be inhabited by wise men who have direct contact with God, and when the Agora's World Lord prays, everything that is alive dies out of terror. The Agarti underground was visited by people whose lives have left their mark on the fate of all civilization: Buddha, Paspa, Baber and Issa (Isza) - Jesus called Christ, who spent his young years there searching for wisdom. Mikołaj Notowicz, Sri Swami, wrote about it Abhendanda and Mikołaj Roelich. (=> Anthology - "Unknown Life of Jesus", Zakopane 1993) There are many indications that in fact - somewhere on the borderline of today's Tibet, India, Nepal and Pakistan - there was a powerful religious center, radiating knowledge all over the world and exerting invisibility on it, but a noticeable effect, as in the novels of Montyhert and Redfield (=> F. Montyhert - "Atlantis and Agharta", Warsaw 1983, M. Redfield - "Heavenly Prophecy", "The Tenth Initiation", "The Secret of Shamballa", Warsaw 1993-2002).

Ossendowski was not there because in his wanderings, he only went to Lhassy then he went to the USA, China and Japan, where in a few weeks, a manuscript of his book was written. The relationship about the state of Agharty has it from the second and third hand, and there are many indications that it is a story about the real world that is surrounded by the mythical and mystical veil of the Great Secret.

FA Ossendowski wrote about the entries into the underground world of Agharta located in the area of the southern shore of Lake Baikal or in the Arshanov mountain range, which is located between Baikal and Mongolia. The second location is according to the stories quoted by him, somewhere on the border between Afghanistan, Pakistan, Tajik and Chinese. This is a huge area, mountainous and desert. Finding in it a system of caves and secret entrances is synonymous with finding a needle in a haystack. In this area, even the most sophisticated technique fails, as Russians and Americans have learned during the First and Second Afghan Wars.

During the activities associated with the liquidation of the bases of Taliban terrorists in Afghanistan, during the siege of the rocky fortress in the Tora Bora massif, the attacking Americans stumbled upon a completely unknown system of corridors and underground tunnels (except the local ones). In these tunnels, the Taliban disappeared and the Americans failed to locate and liquidate them. The action against the Taliban was discontinued, allegedly at the request of the Afghan government, and this is due to the sanctity of these places. And I would go over the agenda if it was not for the information that the famous Polish explorer and a member of the New York Explorers Club - Maciej Kuczyński. (=> M. Kuczyński - "Szambhalla in Afghanistan?" In "Gwiazdy mowią" No. 12/2003).

So, assuming that this American data is real, we have come across something truly unusual and mysterious, which can disappoint us even to Agharta! It is possible that the Moon Cave is part of the tunnel system that entangles our planet, which actually belongs to Shambhalli-Agharty or non-mythical Interterry. In close connection with them remain mysterious tunnels in Babia Gora and its Polish branches, which persistent legends circulate on both sides of the border.

The Slovakian megalithic sandstone balls from the Slovak Kysuce should be explored. Recently, their traces have also been found in Poland, in the vicinity of Węgierska Górka and Beskid Mały. Still, no one solved this puzzle.

Similarly, how is the mystery of the Wall of Giants that is not solved, which stretches from Banská Štiavnica to the village of Dudince, and which is supposedly visible only from above, from the satellites? However, the information about him comes from the Americans, and as has been said here once that it is the least reliable of sources. But we do not reject it because that's where the solution too many mysteries can be hidden.

Another mystery to be explained is the secret of tunnels under the Pyramid in Bosnia, which was discovered two years ago. Investigations also require huge burial mounds in Poland, located in the vicinity of Kazimierz Wielki, and the volume of only one of them is 1.5 times greater than the Great Pyramid of Giza! And they were erected, as archaeologists calculate, around 6000 BC - ie 8,000 years ago!

Besides, thanks to the help of the Russians from the magazine "Kalejdoskop NLO", and especially Ed. Wadima K. Ilin from St. Petersburg, we obtained a lot of information about the amazing cave systems in the Butentau massif in Central Asia and tunnels connecting the caverns of the northern and southern Urals. It seems that what we know about the caves of this world is just a fraction of this mystery, which also includes the mystery of the Moon Cave.

CHAPTER I

Oblazowa Cave

INDIRECT PROOF

Looking for evidence of the authenticity of the relationship of Dr. Horak, we asked ourselves the very possibility of such a cave being unchanged for several or even tens of thousands of years. Our fears were justified because the cave is located in the penseism area of the Slovak-Polish borderland, which means that it is haunted from time to time by earthquakes: sometimes even very strong. Just how much these fears are justified is shown in the table below, which shows stronger earthquakes from the tenth to the first years of the twenty-first century:

EARTHQUAKES IN POLAND

According to the "Catalog of earthquakes in Poland in the years 1000-1970", Warsaw 1972 and author materials.

Date	Epicenter Earthquake	Division	Degree
008 VII [3]	Czech Republic	Moravia Silesia	very strong [4]

---

[3] The earthquake was repeated for 8 days.

[4] Due to the lack of accurate measurements of earthquakes, I give the strength

	Germany		
1000.03.29	Ljubljana (SLO)	All of Europe	9
1011	Kowary	Karkonosze	strong
1014.11.18.	Silesia	Silesia and Poland	strong
1016	Krakow	Poland	strong
1034.02.17.	Poland	Poland, the Czech Republic, Hungary	strong
1040.12.25.	Hungary	Poland, the Czech Republic, Hungary	strong
1044	Poland	Poland, Czech Republic, Hungary	strong
1092.06.26.	Morawy (CZ)	Poland, Czech Republic, Hungary	strong
1170.04.01.	Hungary	Poland, Czech Republic, Austria, Hungary, Ukraine, Germany, Switzerland	very strong
1201.05.04.	Styria (A)	Austria, the Czech Republic, Germany, Poland	9
1258.02.07.	Poland	Poland, the Czech Republic, Russia	strong
1259.01.31.	Krakow Poland	Czech Republic	9
1303.08.08.	Krakow Poland	Germany	strong
1328.08.04.	Hungary Poland	Hungary, Czech Republic	strong
1356	Basel (CZ)	Czech Republic, Poland	very strong
1358	Moravia (CZ)	Czech Republic, Poland	strong
1372.06.01.	Sundgau (D)	Germany, Poland, the Czech Republic,	9

based on a descriptive scale: weak, strong or very strong.

		Switzerland	
1384.12.24.	Austria	Germany, Poland	strong
1433	Lower Silesia	Poland, the Czech Republic, Austria	strong
1441	Slovakia	Slovakia, Poland	9
1443.05.29.	Hungary	Hungary, Slovakia, Czech Republic, Poland, Austria	<9
1443.06.05.	Central Silesia	Europe	9
1443.06.05.	Zvolen n. Hronom (SK)	Slovakia, Poland	9
1483	Brzeg	Poland	weak
1496.06.23	Nysa	Poland	weak
1517.09.25.	Podole (UA)	Poland, Ukraine	weak
1528	Poland	Hungary, Ukraine	strong
1562.02.10.	Kłodzko	Poland, Czech Republic	strong
1572.01.06.	Warmia and Masuria	Poland	very strong
1572.01.09.	Torun	Poland	strong
1590.09.15.	Alps (A)	Austria, Germany, Poland, Czech Republic, Switzerland, Slovakia, Hungary	9
1591.IV-V	Wieliczka	Poland	strong
1591.05.09.	Moravy (CZ)	Czech Republic, Poland	very strong
1601.01.06.	Warmia and Masuria	Poland, Denmark, Sweden	9
1606	Tuczno	Poland	5
1615.02.13.	Kłodzka Valley	Poland	weak
1650.04.14.	Stara Lubovnia (SK)	Slovakia, Poland	4

1594.09.15.	Złotoryja	Poland	weak
1662.08.09[5]	Spisz (SK)	Slovakia[6], Poland	very strong
1671.12.28.	Rzeszow	Poland	weak
1680	Poland	Poland	strong
1690.12.04.	Carinthia (A)	Austria, Poland, Czech Republic, Germany, Switzerland	9
1695.VIII	Hungary	Poland, Hungary, Slovakia	strong
1715.05.01.[7]	Cieszyn	Poland, Czech Republic	strong
1716	High Tatras	Slovakia	5
1717.03.11.	Pieniny	Poland, Slovakia	strong
1724.01.29.	Kežmarok (SK)	Slovakia, Poland	7
1751.06.31.	Karkonosze	Poland	weak
1763.06.28.	Slovakia[8]	Slovakia, Poland, Czech Republic	9-10
1768.02.27.	Central Austria	Europe	8
1774.01.26/27.	Upper Silesia	Poland	> 7
1775.01.24.	Wroclaw	Poland	weak
1778	Low Beskid	Polska	4
1778.05.10.	Lower Silesia	Poland	weak
1785.02.07	Morawy (CZ)	Czech Republic,	strong

[5] According to other sources, it was on August 6th.

[6] An earthquake could have caused a meteorite impulse that blasted the top dome Slavkovský štít, then the highest peak of the Tatras measuring around 2.700 – 2.800 m above sea level.

[7] The earthquake lasted 36 hours.

[8] The earthquake destroyed all the stone houses in Slovakia.

		Poland	
1785.08.22.	Silesian Beskid	Poland, Czech Republic, Slovakia	6.5
1786.01.03.	Szczecin	Poland, Germany	weak
1786.02.10.	Mysłowice	Poland	weak
1786.02.13.	Kłodzka Valley	Poland, Czech Republic, Slovakia	5-5.5
1786.02.26.	Silesian Beskid	Poland	3-3.5
1786.02.27	Kietrz	Poland, Czech Republic, Slovakia	6-6.5
1786.02.27.	Opava (CZ)	Poland, Czech Republic, Slovakia, Austria	7.5
1786.03.04.	Kłodzka Valley	Poland	3-3.5
1786.10.03.	Cieszyn	Poland, Czech Republic	3-3.5
1786.12.03.	Myślenice	Poland, Ukraine, Czech Republic, Slovakia	7.5-8
1789.12.11.	Karkonosze	Poland	> 4
1790.03.13.	Wroclaw	Poland	strong
1790.04.06.	Transs. (RO)	Poland, Romania, Ukraine, Hungary, Slovakia	strong
1799.II	Wroclaw	Poland	weak
1799.IX-X	Jeleniogórska Valley	Poland	weak
1799.12.11.	Trutnov (CZ)	Czech Republic, Poland	6
1802.10.26.	Bucharest (RO)	Romania, Hungary, Slovakia, Poland, Czech Republic, Ukraine	strong
1803.01.08.	Białystok	Poland, Lithuania	strong

1817.02.07.	Red Monastery (SK)	Slovakia, Poland	4
1823.09.09.	Głubczyce	Poland	weak
1829.06.02/03.	Śnieżka	Poland, Czech Republic	weak
1834.02.02.	Silesia	Poland	weak
1834.10.15.	Hungary	Hungary, Slovakia, Poland, Ukraine	7
1837.03.14.	Central Alps	Europe	7
1838.02.08/09	Dukla	Poland	weak
1840.04.25.	Spiska Stara Wieś (SK)	Slovakia, Poland	7
1840.04.30.	Spiska Stara Wieś (SK)	Slovakia, Poland	weak
1841.04.30.	Hungary	Hungary, Slovakia, Poland	weak
1842.02.24.	Cracow	Poland	weak
1842.03.08.	Cracow	Poland	weak
1855.01.25.	Cieszyn	Poland	weak
1857.12.16/17	Old Sącz	Poland	weak
1858.01.15.	Žilina (SK)	Slovakia, Poland, Czech Republic	9
1872.03.06.	Gera (D)	Germany, Poland, Czech Republic	8
1872.12.26.	Bielsko-Biała	Poland	weak
1876.07.12.	Czech Cieszyn (CZ)	Czech Republic, Poland	weak
1877.10.05.	Jizera Mountains	Poland, Czech Republic	weak
1877.11.25.	Kłodzka Valley	Poland, Czech Republic	weak
1883.01.31.	Úpa Valley (CZ)	Czech Republic,	6.5

		Poland	
1892.10.27.	Morawy (CZ)	Czech Republic, Poland	weak
1895.04.14.	Ljubljana (SLO)	Central Europe	6
1895.06.11.	Ząbkowice Śląskie	Lower Silesia	7
1901.01.10.	Úpa Valley (CZ)	Czech Republic, Poland	7
1901.10.21.	Spiska Stara Wieś (SK)	Slovakia, Poland	6-7
1903.08.28.	Černy Důl (CZ)	Czech Republic, Poland	3
1908.05.13.	Cista near Vrchlabi (CZ)	Czech Republic, Poland	4
1908.12.30.	Gołdap	Poland, Russia	3-4[9]
1909.02.11.	Kołobrzeg	Poland	4-5
1909.02.12.	Kołobrzeg	Poland	4-5
1909.05.06.	Krynica	Poland, Slovakia	3-4
1909.11.05.	Śnieżnik	Poland, Czech Republic	5
1912.12.01.	Smołdzino	Polska	3-4
1928.06.11.	Kołobrzeg	Poland	weak
1931.03.30.	Opava (CZ)	Czech Republic, Poland	6
1932.II-III	Poland[10]	Poland	4-5
1934.09.03.	Opava (CZ)	Czech Republic, Poland	4.5
1935.03.23.	Czarny Dunajec	Poland	5-6
1935.07.24.	Moravia (CZ)	Czech Republic, Poland	5.5

---

[9] A series of shakes.

[10] A swarm of underground tremors throughout the lowland area of NE and central Poland.

1962.06.20.	Dąbrowa Górnicza	Poland	3.8
1966.03.11.	Dzianisz	Poland	4
1980.11.29.	Bełchatów	Poland	4.5
1992.06.29.	Kiczera near Krynica	Poland, Slovakia	6
1993.03.01.	Low Beskid	Poland	6-7
1994.06.01.	Augustów	Poland	> 6
1995.09.11.	Domański Wierch (SK)	Slovakia, Poland	5.2
1995.10.13.	Białka Valley	Poland	5
2004.09.21.	Kaliningrad (RUS)	Russia, Poland, Lithuania	5.2
2004.11.30.	Skrzypne	Poland	4.7
2004.12.02.	Czarny Dunajec	Poland	3.3

Of course, earthquakes from the north of Poland did not cover the Polish-Slovak borderline, but the rest could have a significant impact on the condition of the caves in the Tatra Mountains and other mountain ranges in Poland and Slovakia. Visiting the caves of the Western Tatras and the Belianske Tatras after the earthquakes of the 90s of the 20th century, we asked the guides and administrators about the effects of earthquakes on their condition. The answer was always the same - changes that threaten the existence of these caves were not found! And yet, these quakes were relatively strong, as can be seen in the above-mentioned combination, so that some cracks in the walls, underground water sprays or dikes and roof outfits could be created in them. Nothing like this has been recorded.

Therefore, this allows us to make an optimistic conclusion - if the Moon Cave does exist, it should continue to exist, despite the fears of Patrick Moncelet. Moncelet claims that if this cave - actually the Moon Shaft - is an artificial creature, it was protected against seismic shocks and proof of this is a layer of metal or other resistant material that has been covered with its walls. This protection could not stand and as a result, the entire cave sank. She collapsed like a barn roof covered with too thick a layer of snow. We visited caves on both sides of the border - in Poland and Slovakia. They were different origins and sizes. However, they had a common feature - they contained peaks and rock outcrops, the age of which usually oscillated around 10,000 years!

Ten thousand (10,000) years ago, a cataclysm occurred, which caused the collapse of some cave systems and rock outskirts in others. There were huge boulders Wantul and Babia Góra and - as we think - many other caves in Europe's rocky cliffs. But let's turn to Oblazowa. This cave was famous for certain finds discovered by Polish archaeologists, which was described by editor Beata Zalot in "Tygodnik Podhalański" and on the Internet:

PERFORMANCE REVELATIONS

A puzzle of injured hands

What did the crossed thumbs found in Obłazowa Cave come from before over 30,000 years? What does it have to do with the representations of the human hand in a cave painting discovered in Gargas or Lascaux?

During the initiation of boys and funeral rituals, fragments of fingers were possibly placed in a sacrifice to some mysterious god or was it simply prosaic frostbite?

About sensational discoveries in the Obłazowa cave located in the area The Przełom Białki reserve near Nowa Biala has already been written several times.

It turns out that ongoing research and analysis are bringing more surprises.

Let's remember briefly: Excavation research under the supervision of Dr. Paweł Valde Nowak from the Institute of

Archeology at the Polish Academy of Sciences in Krakow began in 1985 and lasted ten years. The cave was literally sprinkled through a sieve, and a series of archaeological discoveries has become a sensation world rank. The most surprises came from the layer of the osma, representing the Pavlov culture. The whole world has circulated information about the world's oldest boomerang found in Oblazowa made from Mammoth tusk, the earliest tools of mining, a fox-headed amulet polar or two thumbs being the oldest human remains in Poland. Radiocarbon studies determined that the age of all these items are around 30,000 years BC.

Paleolithic temple

Initially, scientists made a safe thesis that Obłazowa was a hunting camp. Tedious laboratory tests analyzes uncovered objects and its arrangements have brought about further sensations. Boomerangs used in everyday hunting were wooden, so the one from Obłazowa had to be of great value to an early man as it was a mammoth tusk, in addition, he had an ornament. Also, other items found there made of rare raw materials - a Jurassic shell and a crystal skyscraper Gorski had to be extremely unusual for an early man valuable and rare.

The symbolism of life and death is evidence that it was a special place - a placefor ritual, something like a sanctuary of a man from around 30.000 years ago.

Did they cut or bend?

Recently, the head of the research, Dr. Paweł Valde Nowak, took care of two human bones found in a cave. Although the cave was flooded through sieve, no other elements of human skeletons were found. One of the phalanges of the left thumb were folded in the center of the system in the middle other unusual items, most likely serving in rituals. A little further, there was another human thumb.

In Paleolithic cave painting, there is often a motif human hand, in many cases with missing fragments of fingers.

The most famous place is Gargas, where 231 hands were discovered, of which over one hundred were missing parts of the fingers. Scientists for years argue that these hands have been crippled or that some of the fingers have been used in the sacrifice or the hand was only bent, which could have been for the time being meant to be a sign.

Also, in the Paleolithic skeletons of the tombs in Murzak-Koba in the Crimea, the palm phalanges were missing. It was custom to cut off parts of the fingers during various initiations and funeral rites.

"All started to put together a certainty," says Paweł Valde Nowak. "We did not understand these details at first. Now I think that found in the cave Obłazowa, the thumbs can be the key to the whole puzzle. We can talk again about the next interpretation to find the essence of this deposit."

Obłazowa and Gargas

Until now, scientists have been met with missing fingers in cave painting. Several dozen of such Paleolithic sanctuaries were discovered, some of them dated just like the thumbs from Obłazowa for about 30 thousand BC. "Never before has such a rich and rich been discovered valuable inventory like in Obłazowa. The opposite situation arose. We have objects and fragments of truncated fingers that testify to the uniqueness of this place, the rituals performed here, but we did not find any drawings on the rocks. There are paintings in the other caves, but there are no objects. Perhaps further research and analysis by Obłazowa will help us understand the importance of the injured hands of Gargas," says the lead research scientist.

Further questions are raised by the location of Obłazowa. Pavlovian culture flourished in Moravia, where archaeologists have discovered many everyday places. In the vicinity of Obłazowa, in a radius of about a hundred kilometers, nocluster was found. Maybe it just is not there. Maybe Obłazowa was a remote place of worship, to which an early man traveled as a pilgrimage and stayed here to participate in some important ceremonies for him. What surprises Oblazowa will bring, is not known. Research laboratories are still underway. There is a monograph on the completion cave. Everything indicates that archaeologists will return to research in the cave itself. For now, however, it is impossible due to issues of safety. To enter there, the grotto must be properly secured and prepared. There is no money for this yet. In the future, a cave could be made available for sightseeing. He could stay there and reconstruct the arrangement of objects discovered during excavations. Maybe - as the French do, for example - create a copy of the cave. It is, however, already a field for local authorities. So much red. Beata Zalot. Another journalist, Ed. Anna Bielecka from "Fakt" gives a bit on the subject of the boomerang found there:

Bu-Mer-Ang 11 is "recurring" in the Aboriginal language from Australia. However, Australians were not the first to use this weapon. The oldest boomerang has 30,000 years, it was found in Poland and is kept in Krakow! Even though the boomerang is known as the hunting weapon of indigenous people Australia, Aborigines, this is not their invention. They defend it before them Ancient Egyptians, Indians from Arizona, Eskimos, ancients used to be inhabitants of New Hybrids. In the treasury of the Institute of Archeology and Ethnology

Polish Academy of Sciences, scientists keep the oldest preserved boomerang in the world and it was found almost 20 years ago. In 1985, Polish archaeologists have discovered in Obłazowa Cave in Nowa Biała, Spiš and Podhale borderlands, a cluster of valuable ceremonial items of servants living here before 30 thousand years. A curved, polished fragment of the mammoth's blow was among these items.

"I admit that we did not immediately know that we are dealing with a boomerang," says Prof. Paweł Valde-Nowak from the Institute of Archeology and Ethnology of the Polish Academy of Sciences in Krakow. "At first, we thought it was a carefully made dagger. After closer research, we found that because of the appearance (crescent shape, with a flat-cross section) associated with tools used by Aborigines, it must be a boomerang. Australia is the only area where the use of a boomerang has survived to our times.

So, why don't we know anything about it?

"Until 1995, field work in the cave was in progress then at the institute. Only now will we be able to show the world the find. In March next year, the boomerang will go to the exhibition in Paris, then the conservator must decide in which museum he will find himself," says the professor. Since it survived Obrazów, Raj, Niedźwiedzia and other caves older than the Moon Cave, why would not it survive? There is no rational reason to believe that it is different.

CHAPTER II

Silicon Valley from Diluvium?

On the route of our holiday wandering, there are two interesting objects that the reader recommends. The first of them is Bałtowski Jurassic Park, where you can see realistically devoted animal models from our past, and not only dinosaurs. However, near the dinosaur reserve in Bałtów, there is another peculiarity of the Świętokrzyskie region. It is a striped flint mine in the town of Krzemionka - located about 6 km from Bałtów. This mine was created and prospered around 4,500 - 4,000 BC. In fact, it was an entire mining basin, because there are several hundred mine shafts nearby. A few of them were connected by an underground tunnel and this is how the Museum of the Flintstone Mine Museum was created, which we have visited.

The mine is visited with the help of a guide. At the very beginning, he makes a small introductory lecture on the use of striped flint. It turns out that the miners of the Świętokrzyskie Mountains supplied flint practically all of Europe at that time - the tools were found in the Pannonian Plain and Slovakia (scrapers and arrowheads can be seen at the Vychodoslovenskeho Kraj Museum in Košice). Besides, the artifacts were found in distant countries like England or Spain! Really

- the flint was the stone that affected the history of humanity in a significant way! Incidentally, the application and processing of flint was shown by the Czech writer, Eduard Štorch, in his book "Mammoth Hunter" (Warsaw 1952) with the description of an even earlier world, 20,000 years ago (Dyluwium, Pleistocene). According to this author, flint processing was as follows:

"And here is a stone," Kopacz pointed out, and dug a bone around the stone to retrieve it. Żabka helped him eagerly. After a while, a large stone like two male fists flew out of the cave. At the same time, he returned to the cave Kudłacz with a hunted lynx. The thrown stone hit him in the leg. The hunter threw the lynx and picked up the stone. The hunter jumped up and down the cave, tossing the stone into the air, and was not angry at all. It was just the opposite: his face was happy and he shouted cheerfully, as if he was enjoying something. The children ran out of the cavern to watch as the old Kudist danced.

"Flint, flint!" Kudłacz rejoiced, throwing the stone up and grabbing it in his hand. Now they will have flint knives, sharp tips, sharp scrapers, sharp bits and sharp awls. As you can see, not much has changed since that time to the launch of the strip flint. However, flint processing and tool making at that time were as follows:

Only the cluster commander had the first to try to split the flint. Skner took him in his hand, watched him closely and consulted with the most expert hunters on how to smash the flint. When they noticed some slight scratches on the stone, Mamucik explained: "Flint on fire and in water!"

This meant that the flint was once heated in the fire and then thrown into cold water. This activity was probably repeated many times, until the compact flint core cracked. Now it will help them a lot. Sknera took the flint in his hand and moved between the rocks. The hunters followed him at that time, surrounding him with a tight circle. The chief lifted the stone with both hands high above his head and with all his strength, he hit it against the rock. In turn, they threw it to Zabijak, Ukmas and finally Mamucik. Finally, the stone broke into three pieces. Sknera's chief, Kudłacz and Mamucik, took a piece, sat on the rocky ridge and flattened the sharp tiles with flint. In the skilful rinsing, the thin blades of knives and the angular translucent edges of the pyramids, which could serve as bone and skin scrapers perfectly, flaked off the flint. (E. Štorch - "Mammal hunters" - op cit. Pp. 104-106)

This was the case in 20.000 - 30.000 BC. Methods of flint processing have not changed much, but the thing is that in the area of Ostrowiec Świętokrzyski 6000 years ago, the industrial production of striped flint tools and jewelry was in full swing! And then came the metal age.

We went into the underground. Mine tunnels run about 12 m underground. They are carved in white, limestone rock and the ceiling from the floor divides about 2 meters, so you can walk there without bending down. The miners of that time had to squeeze in corridors and adit spaced 50 x 50 cm. The works were carried out mainly in winter. The corridors have a constant humidity and temperature of + 7°C. It was silent and chilly and darkness at that time was illuminated only by the pale lights of the lamps.

After exploiting one shaft, it was filled with spoil - a white limestone rock. The right raw material - powerful flint rolls weighing a few kilograms - were divided into smaller parts, which formed the basis for the production of proper knife blades and axes, arrowheads, leather scrapers and ornaments. By the way, I wonder if there were any iron and stone meteorites in those deposits of alkaline rocks that fell when these settlements were formed, and so in Trias? If there was such a good search, there would be more than one! Who knows if such finds did not inspire our Pra-pra-great-ancestors for smelting iron? After all, there is an old Polish metallurgical center, in which iron was smelted using a homemade method, using a smoke machine. This method is primitive but effective and not so long ago used in China!

In addition, let us not forget that in the area of the Świętokrzyskie region, there are also iron ore deposits (chalcopyrite - $CuFeS_2$, pyrite - FeS_2, siderite - Fe $[CO_3]$, rarely volcanoes - Ca10 (Mg, Fe) $2Al4$ $[(OH)\ 4]\ (SiO_4)\ 3\ |\ [(Si2O7)\ 2]$ and turf ore: limonite - FeO (OH)), copper deposits (native copper - Cu, covellite - CuS, kuprone - Cu_2O, azurite - Cu3 $[OH\ |\ CO_3]$ 2, malachite - Cu2 $[OH\ |\ CO_3]$, chrysocolla - Cu $[SiO_3]$ · nH_2O, konichalcite - CaCu $[OH\ |\ AsO_4]$). And uranium!

Apart from the Sudetes and the Suwałki Region, the presence of uranium was only found in the Świętokrzyskie Mountains. In the years 1958-1965, the Department of Rare and Radioactive Elements carried out, in this region, a study on the uranoceneity of Paleozoic and Mesozoic sediments. There were many anomalies and several uranium mineralization points. The most important uranium deposits in the Świętokrzyskie Mountains include the areas of Rudki, Miedziana Góra, Miedzianka, Daleszyce and Winna. Instances with clearly increased uranium content are nested and always associated with Variscan dislocations. Minor extraction was carried out only in Rudka near Nowa Słupia. (=> W. Rejman - "Uranium mines in Poland" in "Wiedza i Życie" No. 9/1996). In this region, there is autunit - CaO (UO3) P_2O_5 · $12H_2O$, pitchblenda - UO_2, carnolite - K2 (UO_2) 2 (VO_4) 2 · $3H_2O$ and composites - Mg (UO_2) 2 $(HSiO_4)$ 2 · $5H_2O$. (IEA, lecture "The Fuel Cycle in Nuclear Energy" - http://www.iea.cyf.gov.pl/rysunki/energetyka/energetyka_03.pp)

By the way, uranium ores in this region of Poland were discovered by the Russians, while the Germans, during the Second World War, intended to exploit them and perhaps use them to build their atomic bombs and other "nuclear devices".

All I am wondering is WHO, WHEN and HOW did he point out to people that it is here that one should dig up this valuable (for them) resource that decided their survival? Who do these people owe? For me, this is a "Daeniken miracle", and Humanity survived in struggling with the hard laws of Nature. However, there is a fundamental difference between the views of Erich von Däniken and mine. Von Däniken claims that this is the knowledge Humanity has received from our Space Brothers and that has been applied by people in their daily lives. I think that this is the knowledge acquired through many years of experience, which was a sine qua non condition for Humanity's survival after the Atlantean catastrophe.

Well, there were no aliens. It was, however, equal to ours and perhaps a little superior to the civilization of Atlantis, which died as a result of not one, according to Plato (Aristocles), but two cataclysms. Other authors, such as Sir Brinsley le Poer-Trench, claim that this was a whole series of cataclysms, the last of which caused the sinking of Poseidia - the last island of Atlantis. The first of them was the shift of the plume of hot magma from Atlantis and its sinking as a result of lowering the bottom of the Atlantic, and the second was the collision of a comet with Earth over North America. Both of these events took place shortly after 13,000 years ago and Humanity's survivors had to cope somehow. These catastrophes brought them back to the state of a stone not yet broken, and as a result, only legend remained after the magnificent civilization. I refer to those interested in Ludwik Zajdler's perfect monograph - Atlantis. People had to start all over again. Knowledge about the world and the Universe gradually became a collection of religious myths and has survived to this day as different religions of the world. It has not only left memories, but also specific information useful for the descendants of Atlanteans were left. We understand some of them only today, in the 21st century. In this context, the paranoid interest of the Nazi "top-hat" of knowledge and religions of the East is understandable. Simple, specific messages from the past were sought for obtaining new energy sources. New weapons and means of mass destruction constituted the margin of this search and energy was the most important! Without her, the Third Reich could not lead a war and the conquest of the world, which was assumed by her insane ideology.

But let's go back to Krzemionki. Impressions are unforgettable, especially in the form of miners working with primitive tools from wood, stone and animal bones. Dark flint rolls protrude from the white walls and some of them resemble hubs or oysters. I do not want to believe that 6,000 years ago, people were able to cope with such problems as shuttering walls and reinforcing them with artificial pillars instead of wooden ones. This was because they broke under the weight of the overburden, but with intricately arranged stones, they were very durable and relatively safe.

We really have nothing to be ashamed of because our ancestors were no worse than the builders of pyramids or other megalithic buildings that are the pride of other nations. What's more, our mines supplied tools for people all over the continent, and it was thanks to them that they could reach the first scientific and technical revolution.

It may sound pretentious, but this is the truth. Krzemionki and other towns where the striped flint was mined were the Silicon Valley of the time, which fueled progress during the dawning of our civilization.

When we visited the flint mine, it associated us with a crystal grotto from Mexico. Do you remember the drawings for the book, "Journey to the Inside of the Earth" by Jules Verne, when Prof. Lidenbrock, his nephew Axel and guide Hans Bjelke are strolling around the underground world, among the great crystals of various minerals. It turns out that Lord Julius foresaw such a possibility.

Since autumn 2001, works have been underway to investigate and describe one of the most uncommon discoveries in the history of speleology. It is a huge crystal cave called the Dream Cave, which was discovered in Naica mine located in southern Chihuahua in Mexico. These huge crystals exceed all acceptable standards known to geologists. It is a huge emptiness in the geode shape, filled with crystals of 10-12 m in length and 1 - 1.2 m in thickness. They shine silvery and gold in the light of lamps and their mass reaches up to 10 tons. The cave was discovered in a limestone massif in which there are also deposits of silver-zinc-lead ores. The material from which the gypsum was most probably formed - $CaSO4 \cdot 2H2O$ and selenite - a fibrous variety of the same plaster. And this reminds me of the applications, according to which the Atlanteans were to use huge crystals for obtaining energy. Maybe in massive crystalline lasers? Gypsum is not gypsum but the huge crystals arouse respect and amazement - in the world, there are only two caves with such huge crystals. Maybe they have something to do with Atlantis??? Details can be found on Richard Fisher's website, who researched these caves: http://www.canyonsworldwide.org/canyonsorg/books.html. Or maybe it was the Mexican version of the old Polish Silicon Valley? Who knows?

Wandering the underground worlds of Kielce, we mentioned another mystery of the past. It is a mystery to the Butentau cave complex. The article by Vladimir Niecziporenka on the subject appeared on the pages of the Russian weekly, "Kaleidoscope NLO" No. 35/2007 dated August 27, 2007. He mentions a strange land that is located in Central Asia, and perhaps the fact that there is a source of legend about the saint Śambhalli (Szamballi) and at the same time, shows what our ancestors were capable of.

High caves

In the dead and uninhabited region of Central Asia, south of the Aral Sea, there is a mysterious land called Butentau (Buten-tau). It is here that there is a dry and lifeless Ustjurt (Ust'-yurt) Plateau falling through rocky steppes and rocky towers to the sandy dunes and deserts of Kara-kum.

At some point, the spas form a rocky precipice high at 70 m and stretching for tens of kilometers. Next to it, in the midst of the dunes, one can see traces of the existence of a huge, ancient canal on the banks of which rose the massive ruins of the early medieval fortress of Adak. In the upper part of the yellow-gray rock walls, at a height of 50 m from the foot, perfectly black spots of holes are visible. They are the entrances to the caves and there are hundreds of them!

Judging from their construction, these caverns and caves are not natural creatures that arose as a result of erosion - they have artificial origin. Yes, how else could it be in a steep, overhanging wall, with a height of over 70 meters?

Even before the war, in the area of the Adak fortress, he conducted excavations known to the Soviet archaeologist and ethnographer, Prof. Siergiej Pawłowicz Tolstow from the MGU. Unfortunately, he was unable to examine these caverns, mainly due to the lack of specialists in this field and technical means to work at such heights.

It was not until the 1970s that various history buffs came to Butentau from such cities as Kiev, Odessa, Kharkov and Uzhgorod. After several seasons, they examined about one hundred cavities using special techniques. Later, local enthusiasts from Tashkent, Samarkand, Bukhara and Urgench joined them.

Among the participants of one of such expeditions was a resident of Tashkent - Nikolai Alexeyevich Golubev, who later moved to St. Petersburg. The laconic narrative of this researcher points to the riddle: Butentau is hidden deeper than we think.

Drawings on the walls

"The caverns were similar to one another, like rooms of one flat," says Gołubiew. "Their depth rarely exceeded ten meters. We came to the conclusion that these caverns were used for living because we found chimneys, beds and everyday objects. The walls of the caverns were covered with petroglyphs depicting domestic and wild animals, scenes from hunting, fighting and warriors. These beautiful drawings still breathed life and could be a great exhibition for many museums. We also solved the mystery of how people entered their high-rise residences Ladders made of large animal bones have been found in some of the apartments. Apparently, the inhabitants of the cavern had rope ladders that pulled into the interior overnight. It can be assumed, therefore, that people hid from the invasion of enemies in the caverns.

Our expedition was coming to an end when something happened, about which I was silent up to now, and for which I did not find any explanation. You will understand why."

Disappearing tunnel

Our group, consisting of three people, was going to the next cavern. On this fragment of the wall, a little lower and sideways - at your fingertips - the entrance opening to another, unexplored cave was blackened. In appearance, it did not stand out from the neighboring ones. For now, I did not intend to look there but when lowering myself and I reached the level of this cave, I heard from its depths a strange sound like I have never heard in this place.

Curious, I climbed to the edge of the cave and entered the hole. I sat with my eyes closed for a few seconds to get used to the darkness and when I opened my eyes, I realized that this cavern did not have a back wall! The entrance was sloping and led to some flat space that had no end! And this passage or rather the tunnel, was not dark but the air in it was shining with a pale yellow light! I entered the tunnel and noticed that it was gradually getting wider and higher. It seemed to me that I was going for quite a long time but in reality, I walked about 50 meters. But that was not a problem as I have already understood that this corridor is very, very far away. Suddenly, voices came to my ears. I thought that it was my colleagues calling me, who in the meantime dropped to the ground and cried to me after noticing my absence. I called out to them to share my joy and confirm my sensational discovery!

I returned to the entrance and leaned my head out of the tunnel-cave opening. "Hi!" I shouted.

I did not cry out anymore because behind me, I heard this strange sound again, like moving stones. I turned away quickly. At a distance of about 10 meters, a rock slab was lying on the floor, which closed the tunnel from which I had just left. Half a minute and the tunnel disappeared! Literally, as if it never was! A wall was now in front of me, as rough and monolithic as the walls in the other caverns.

"What happened?" my companions asked worriedly.

"Nothing. I will drop off to you soon," I replied. I ran to the wall and lit it with a flashlight. I did not even find a crevice on its periphery. After five minutes, I was on the ground and I did not tell anyone about my discovery. Well, because I did not even have the slightest proof of the truth of my words, and I did not want to be considered a boaster and a fantasy. (In scientific circles, such opinion is a civilian death for a scientist.) Nevertheless, I remembered the place where this cavern was located and decided to return to it next season, try to enter the tunnel again, make documentation and try to penetrate deep into the mystery of Butentau. Unfortunately, for reasons beyond my control, this expedition was the last expedition for me.

Speculation

"In short, I have two hypotheses on this subject," says Gołubiew at the end. The first one is historic. In time immemorial, when Amu-Daria flowed into the Caspian Sea, Butentau was a fertile, flourishing land. It was here that life was brewing, crafts developed, science flourished, and merchants came from all over the former Orient. Also, the various conquerors did not pass indifferently to these fertile and rich lands. Well, apparently, the local rulers decided to build some shelter for the inhabitants against the invasion of these conquerors. Thus, numerous caves/tunnels were carved in solid rocks with entrances that were masked with moving stone slabs: moved by clever mechanisms. Apparently, when I was climbing the wall, I pressed a performance, which in turn activated the mechanism which, despite centuries, still retained its ability to work. Isn't that strange? No similar mechanisms work, for example, in the Egyptian pyramids from the time of the pharaohs. Well then, the same mechanism returned the disc in the same location. Fortunately, at the moment I was not inside the tunnel.

"The other hypothesis is one in which I present a version not from this Earth. As you know, interiors of Central Asia for some reason are in the interest of UFO pilots. Their flying ships were often seen by the inhabitants of even the most distant Kiszłak and Auł, while these UFOs were flying out of the Earth.

"On the other hand, in the mythology of the nations inhabiting it, people often talk about the populated interior of the Earth and the human beings living in it, who can do miracles. As you know, myths and legends do not appear out of nothing, and their origins are usually real historical events. Perhaps I just appeared in this cave at the moment when the Przybysze "pontiled" their underground base???

The thing needs some comment. I realize that in my flawed comment, I will not be able to exhaust this topic in any way. I would like to draw attention to only one aspect of the matter, namely that people - especially the primeval ones - lived in caves, and hence they were given the name of cavemen. Probably because the cave apartments were the best because of weather conditions, defensive capabilities, etc. Yet, there is evidence that it was not. For example, in Ukraine, huts made of wood and punches and other bones of mammoths, hairy rhinoceros and other large mammals were built. Wandering hordes of primitive people lived in portable tents made of animal skins. People settled in houses made of clay and animal manure, cut from reed and other materials. The caves were lived in rather occasionally and yet, entire cities of underground dwellings were built - that I will mention only the famous underground city complex in Cappadocia (Taurus Mountains) in Turkey and the Easter Island. Now it's Butentau and it's been mentioned so many times by different authors of Aghart. No, I do not believe that Agharta is a land within the Earth. It also has nothing to do with Shambhalla and its capital, Kalapa. Agharta is nothing more than caverns hollowed out above the lake in the vicinity of the sacred Kajlash Mountain, as I have already written about. It was there that the sages of the Bön religion, earlier than Buddhism and Lamaism, resided.

As for UFOs flying out from under the Earth, legends about such incidents are circulating in all parts of the world. How much is the truth in them - it is difficult to find out. Perhaps these are Inter-terran or Aghartian vehicles, or rare optical phenomena.

Nevertheless, it is worth answering the essential question: why in a certain period of the existence of people on Earth appeared among them a kind of fashion for underground construction, as if something threatened them from the clear sky? Was it a memory of the destruction of Atlantis and atomic wars of the gods-astronauts? The wars that have brought the Earthlings back to the state of primeval wildness?

I think it is worth looking for the answers to these questions. One of them may be found among the hot sands and rocks of Butentau.

There are those like Nikolai Ziatkov, who are not afraid of breaking the title in "Arguments and Facts" with fat Versailles:

WE FOUND THE LEGENDARY GARAGE TO SZAMBALLI.

And then they write like this:

On the pages of our journal, many times we have published materials about expeditions of Prof. Ernst Mussasza in the Himalayas, Tibet and Egypt. This time, the tireless traveler goes on a journey - to a mysterious Easter Island and, which has already become a good tradition, he gave an interview to readers of "Arguments and Facts". Today, he will talk about the goals and tasks of this expedition to Nikolai Ziatkov. This interview was published in "AiF" No. 25 and 26/2004.

Dangerous sign of sixes

Question: Ernest Rifgatowicz, as far as I know, you put this expedition on three occasions, and this time you did not translate it again. What were the previous changes to the expedition date related to?

Answer: There were two reasons for this state of affairs. First of all, by organizing the expedition, I came to the conclusion that from a scientific point of view, we should "see" it and it takes time to "mature" it. Secondly, I would like to point out that the expedition to Easter Island could have been dangerous. It so happened that it would be my sixth expedition (after the Himalayas - 3, Tibet - 1 and Egypt - also 1) and therefore, I did not have much desire to go to this complex and difficult expedition under the sign of the six.

Q. What could be dangerous on Easter Island? This is a small island in the Pacific Ocean, where there are stone statues.

A: The point is that the entire Easter Island - having a size slightly larger than 20 km in diameter - is cut by a network of underground walkways and artificial passages that create a labyrinth in total. According to local legends, the underground labyrinth has several branches, each of which leads towards the center of the Earth.

Q: So, you are going to go through this maze - why?

A: On the side of the Earth opposite to Rapa Nui, there is the holy Mount Kailas (Kailas), around which lies the legendary City of Gods, which consists of many huge and very old pyramids. This city of Gods is described in my last book, "In Search of the City of Gods. Volume 3. In the embrace of Shambhala." And it was in the Tibetan City of Gods that we found the legendary Gate to Shambhala - a leading and legendary, underground city - the capital of underground Shambhala.

City of Gods

I cannot rule out the fact that on the other side of the globe, i.e. on Easter Island, there was the second City of Gods - even older, which sank in the ocean. The underground city - taking on logic - should also be there and the entrance to it is located in the underground labyrinth of Easter Island.

Q. When you speak about the City of Gods, wanting or not wanting, a question is born: what is it?

A: It is difficult to answer this question briefly. This is what my next book will be dedicated to, "In Search of the City of Gods. Matrix of Life on Earth." The thing is, when we delineated the map of the City of Gods, it turned out that it has a structure very similar to the structure of DNA. The well-known Russian molecular biologist Dr. Piotr Pietrowicz Garjajew has just stated this.

Q: Huge stone? DNA?

A: Yes. We are under the conviction that the City of Gods is the place where God created man on Earth, manipulating DNA first and foremost.

Q: And Darwin's theory?

A: It's just ridiculous.

Q : Did you say that there are two Cities of Gods?

A: I believe that the first city of the Gods was a city sunk in the area of Easter Island, and the second - a city in the region of the holy mountain Kajlash in Tibet. We cannot find the First City of Gods - it sank in the Pacific Ocean.

Q: In which case, why are you undertaking an Easter Island expedition, since the Number 1 City of Gods has sunk in the waters of the Pacific? For what purpose do you want to explore the underground Rapa Nui maze?

A: The whole thing in the legendary stone of Shantamani (Czantamani). Our research during the Tibetan expedition in the area of the City of Gods led us to the conclusion that this stone is in the pyramid of Little Kajlash. This is located on the western slope of the sacred Kajlash Mountain and standing in an extremely inaccessible place on three stone columns with an approximate height of 600 m. We assumed that the Shantamani stone is a "stone plaque" on which the program of the origin of life on Earth was written - many Tibetan legends say this about it. Well, once you think about the fact that there are two Cities of Gods, then in the City of Gods in the area of Easter Island there should also be a corresponding stone, Shantamani - the oldest program of creating man on Earth. And it cannot be ruled out that he is in the basement of this island.

Q: Do you want to find the stone of Shantamani?

A: Human curiosity is not everything. To do so, it needs preliminary analysis to at least find some confirmatory facts! It can be assumed that some old, highly developed people, expecting the City of Gods sooner or later to be sunk in over two kilometers deep, decided to move this "Stone of Life" to the highest point, which turns out to be today Easter Island and hide it there. A talented ufimsky mathematician, Dr. Szamil Cyganow, made calculations of the position of Rapa Nui relative to the place where the axis from the sacred mountain of Kajlash and the center of the Earth is exactly to the other side of the globe. It turned out that Easter Island is exactly 999 km to the west. But the most interesting in all this is that the difference between the length of this axis - from Kajlash to the opposite side of the geoid - equals to 19,999 km, and the axis connecting Kajlash with Easter Island - 19,333 km - is exactly 666 km! According to the equation: 19.999 - 19.333 = 666, the axis carried out from Rapa Nui through the Earth's center leaves it somewhere in the northern part of India - or more precisely, on the river called Chambal (phonetically sounds like Shambhala) - which is a tributary of the holy Ganges.

Q: It means that there are ominous numbers, 999 and 666, which you wrote about in the first volume of his book "In Search of the City of Gods - The Tragic Message of the Olds"?

A: There is some regularity in these numbers. For example, the altitude of the sacred Kajlash Mountain is 6666 m (other sources give 6714 m above sea level - crowd note), and the distance from the Stonehenge complex in England to Kajlash is exactly 6666 km. The distance of the Great Pyramid in Giza, in Egypt, to the North Pole is also 6666 km - and so, nine times only in this quarter of the globe. There is something coded in these numbers. The arrangement of all these monuments is related to the numbers 6 and 9. Thus, Easter Island is a chosen one which cannot be forgotten. I would also like to note that the Pyramid of Little Kajlash in Tibet where, according to our hypothesis, the legendary stone Shantamani is supposed to be located at an altitude of 6,000 m above sea level. Therefore, it can be assumed that the second stone, Shantamani, located on the other side of the globe is at a depth of 6000 m from the ocean level, to which secret underground tunnels and windscreens from Rapa Nui lead. There are no high mountains on Easter Island.

Q: Will you not go down to such depth?

A: Not everything at once. But I think that we will find many interesting things anyway. I would like to live a little longer.

Q: And what are - according to you - big stone statues on Rapa Nui?

A: They may be the proximity of the Shantamani stone. Just like the Tibetan City of Gods, it is marked with the statue of the "Reading Man" high on the 15 floors that we saw, photographed, filmed and which, according to local legends, is a sign of the holy stone of Shantamani.

(These stories were written by, among others, FA Ossendowski, M. Roerich, H. Bławacka and many others). In the pages of "Nieznane Świata", Dr. L. Szposznikowa quotes them in the article "Shamballa dawna i zagadkowa." In addition, I recommend the reader the perfect cycles novel J. Redfield - "Heavenly Prophecy", "The Tenth Initiation", "Secret of Shamballa" and sensational novels of E. Pattison - "Mantra of the skull" and "Water washed stone", which takes place near Saint Kajas - note: RKL)

In addition, I can add that, as they say, the eyes of figures on Easter Island shine in the second decade of September each year. And at this time, we are going there with our expedition.

Q: And the last question is: you are coming out on the Easter Island marked with the numbers 666 and 999, you will be with your 6th expedition.

(NB, Father Sebastian Englert counted that on Rapa Nui there are just 666 stone statues of one type of the Rano Raraku quarry carved in the rocks - attention RKL)

A: No, no. She will be seventh! The sixth expedition will be a second expedition to Egypt. And that's because after adding 666 to the next six, we get 6666. The pyramids and monuments of ancient Egypt, which we want to see for the second time, from another point of view, will be an inspiration for us during the research on Easter Island. How many puzzles are there in Egypt! The complexity of pyramids and other monumental buildings in Egypt is the youngest in the world. Many of them survived, while megalithic monuments on Easter Island vice versa are the oldest ones. Many will give, I think, analogies between Egypt and Rapa Nui. So we will go around the unlucky six during the expedition to this island, which is marked with the numbers 666 and 999.

Q: So, have a good journey and be on your own!

A: Thank you. Everything is subordinate to this Earth, but not to us. Something higher controls us.

From the "AiF" editorial office:

The Mułżeszew expedition is reachable on Easter Island only by means of a satellite telephone; there is no other communication with it. We will, from time to time, post brief information about its progress and the results of research conducted by it in the pages of our journal.

* * *

So, should Shamballa be sought among the scorching countries of the Middle East or among the Himalayan glaciers? Or maybe among the mysterious and picturesque mountains of the Turkish

Cappadocia? Here is the report by Wiktoria Leśniakiewicz about the peculiarities of this land:

The book, "Gods of Atomic Wars" talks about a terrible armed conflict that took place several dozen thousand years ago. Some authors, such as Sir Brinsley le Poer-Trench - Lord of Clancarty - estimate this period of time between 12-50 thousand years ago. [11]

Others for equally distant times, but in these cases, the caesura is here 12,000 years ago or with the breakthrough of the Zodiacal Epochs of the Virgin and Lion - the peak of the Empire of the Atlantic Empire in the Mediterranean, which at the same time became the beginning of its downfall.[12] The remnants of these shocking events have become strange artifacts, megalithic constructions, passages and legends, and just these unbelievable facts. These are surrendered by orthodox science, because they do not fit within its narrow framework and with the chain of such strange facts we met in Turkey in the summer of 2002.

Cappadocia is a geographical region located in the Taurus Mountains, in the south of Turkey. Its landscape resembles a landscape from another planet. It was here that some episodes from the Lucas Star Wars were made, which was created about 60 million years ago in early Cenozoic, shortly after the impetus of the asteroid that killed dinosaurs. This land then belonging to the land of Tethys was covered with a very thick layer of bright volcanic tuff - produced by three nearby volcanoes: Erciyeş Dag (3,917 m above sea level), Hasan Dag (3,263 m) and Gölü Dag (2,143 m) - which then cracks

---

[11] Brinsley le Poer-Trench - Operation Earth, London 1978 - translation by R.K. Leśniakiewicz.

[12] Brinsley le Poer-Trench - "Men Among Mankind", London 1978 - translation by RK Leśniakiewicz; Ludwik Zajdler - "Atlantis", 4th edition, Warsaw 1980.

creating an amazing looking field formations: pinnacles, cones, rock mushrooms, caves, ravines or canyons carved out by rainwater. Amazing and beautiful is a land.

Sometime later, for 1,800 years BC, these areas were inhabited by the Hatti people - the Hittites who created their own empire. The people who settled there then built cities in tuff rock. The thing is that they were cities not on the surface of the Earth, but in its depths and that's what interested us, especially Derinkuyu, which I visited in August 2002. The guides talk about Cappadocia and its underground cities that they are unique on a global scale. In fact, there is no such thing anywhere else or at least not yet found.

In this region, there are 36 underground cities and everything indicates that this is not the end, and that the largest and most interesting discoveries are still far ahead of us. Who can know how much time was spent building these underground complexes and how many thousands of people worked on their construction? Who knows what construction techniques they used, how they extracted thousands of cubic meters of rock and soil, how they transported the surface and how they got rid of it, not to draw the attention of your enemies to the ongoing underground work? Indeed, everything was the result of human work and ingenuity, although it was not easy to do at the then technical level.

The most logical method of building underground complexes - which explains a lot - was drilling mainly ventilation shafts, which were hammered into the rock until the water was found, that is to a depth of 70-85 m; the levels of the underground city were then extended from the ventilation shafts to the depths of the rocks in all directions. The ventilation shafts were always open, so that the underground constructors and builders had fresh air. Thanks to this, a great work of art and design sense of the people of that era was created. How was it done? Probably with stone chisels and axes in a soft tuff. No traces of tuff processing were found, but in 1910, English archaeologist R. Campbell Thomson found such tools in one of the underground chambers.

This question is born immediately and concerns the issue of drilling underground structures, namely: how was the output from the earth mined and what was done with it? Of course, whatever that was chosen from the ground - the soil and rock extracted from a depth of 70-85 m in certain places and all this was done in the area of up to 4 km^2 - would be built into a large hill or mound. It is puzzling whether these cities were built during Paleolithic times. Nobody knows that yet because research and excavations are limited. There is evidence that Romans and Byzantines lived there, and there is a cross-shaped church there, where a later missionary school was established.

Ventilation of underground cities was at a very high level. Clean air reaches the lowest level through a system of tunnels and ventilation shafts. It is shocking that the air is as clean as if filtered. Cigarette smoke is removed impossibly fast through the ventilation shafts on each of the seven floors of Derinkuyu! On individual floors, the temperature in the vicinity of ventilation shafts ranges from +7 ° C to +8 ° C and is constant regardless of the season of the year. In the parts distant from the shafts, it is higher and is between +13 ° C and +15 ° C. The air flows through a system of holes 10 cm in diameter and 3-4 m long drilled with wooden drills with metal ends.

So far, a small part of the underground has been made available to visitors. We do not have much information about the history of these cities. For this reason, we are constantly faced with new questions. For example, we still do not know how people could live in such crowded dungeons. It was said that 200,000 residents lived there once! It's a lot, for those times!

There are not many kitchens on the route available to visitors in Derinkuyu. It would look like there were kitchens for one or two families. At the moment, the view prevails that many families used one kitchen together - as in the former USSR. It still had a strategic and defensive meaning - smoke from many cuisines could reveal the location of underground cities to enemies. Therefore, people used only a few kitchen furnaces that served the entire population.

The next thing is the matter of the toilet. In the underground cities of Tatlarin and Gelveri, toilets have been made extremely carefully and can be used even today! There are even underground cesspools. In other cities, the waste was stored in clay and closed vessels, which were then taken from the vaults to the villages, which masked the entrance to underground cities, and thus prevented the stench and the possibility of infection with salmonellosis or other diseases of this type.

There are no traces of clothing worn by people living in these dungeons - at least they were not found in the chambers and corridors along the tourist trail. Thicker garments were definitely used because of the lower temperature in the basement. The corridors of all underground cities are 160-170 cm high from the floor to the ceiling and it is not clear if the people living there were of that same height.

It is known that there were farm animals and wineries, as stables, cowsheds and wine bars located at all levels of the city indicate. Meanwhile, the villagers cultivated the land on the plains. The land was volcanic and therefore very fertile. Here is another question to ask: how did these people communicate with each other when their enemies attacked?

In Cappadocia, there are hills and mountains such as Erdaş, Karadağ, Çağnı and Kahveci - on the tops of which there were sentinel sentries, which transmitted information on threats with the help of flashes of light and other signals. Typically, in such cases and in such area, smoke signals were communicated.

Underground cities were also important during the spread of Christianity. During the excavation and cleaning works, old coins, ore pots and lamp oil were extracted. However, underground cities have not been inhabited since the eighth century AD. For those 1,200 years, they became open and penetrated by rain and snow, dust, stones and earth through open doors and ventilation shafts. Meanwhile, people still lived in the villages above them, not realizing what was under their feet.

The most interest in modern times is aroused by unusual doors made of round, smooth hewn stones, 55-65 cm thick, 170-175 cm high and 300 to 500 kg weight! They are much harder than the tuff rocks forming the underground city. They were cut and worked on the surface of the earth, and then transported down, where they were permanently embedded.

Generally, all cities - those made available to tourists and those unavailable to the public - are located on the eastern, southern and western slopes of the hills. On the north side, they were not built due to the harsh winters and a lot of snow that fell during them.

So far, no one has answered the basic questions: who first built these underground cities, in which age and what tribes have done this and most importantly, whether or not the zego built them?

Our guide, with some immensely complicated Turkish name that we've translated into Polish Jack, told us that two of these underground cities are connected by a tunnel with a length of 40 km in a fossilized tuff and obsidian. It sounded incredible but we have been reminded of stories from the Polish and Slovak Spisz, which talk about the mysterious tunnel connecting the Dunajec Castle in Niedzica with the alleged Castle of Carpathians (from the novel by Julius Verne titled "Castle in the Carpathians") in Spišské Podhradie near Prešov in Slovakia. This would also have a length of almost 40 km and run in the limestone rocks of Pieniny and Spišsko - Šarišský Medzihoria.[13]

However, there are things that raise our doubts, namely: at which level animals were kept? 1 or 7? At first, it would not be very safe because the roar of cattle could be heard even from under the ground - especially at night. At the last one too because the feces dripping and leaking through the porous rocks would contaminate the water, this fact is becoming a major problem!

It is said that the local population hid in these cities before the Persians. Of course, the procession of the Persians began in the spring and ended in late autumn, and everything was dependent on the spies. Do not move without it. From Persia - today Iran - Cappadocia is almost 2000 km in a straight line. In practice, much more, because you do not go there through the plains, only mountains. Assuming that the Persian army was moving at speeds of up to 20 km a day - which is an over estimate - these two thousand kilometers would take 100 days or three months and then you had to return to the Iranian Upland also for 3 months. - The occupation of these areas remained.

Another doubt - since the Persians occupied these areas in the years 546-333 BC, they had to know about these cities. It is not possible to hide the fact of their existence for a longer time as the supplies stored in the underground will run out. There were fields that needed to be cultivated: plowing, sowing, mowing and gathering, animals that had to be fed, etc. and this could not be concealed from the invaders. There were no such miracles. A "God's" explanation pleases the townspeople, but not someone who is involved in agricultural work!

Before the Hittites, that is 1,800 years ago, these areas were inhabited by Asians, but they did not build these underground cities.

---

[13] See Robert K. Leśniakiewicz - "Project Tatry", Cracow 2002.

They were created in the era of the early Paleolithic - about 1 million - 100,000 years ago! The question is - to whom and what were such cities needed for? So far, official science has not found an answer to them, and yet the answer is simple, when we look at the problem from the point of view of the hypothesis of the ancient Civilization on Earth - a civilization that precedes our own and whose fall ended the destruction of Atlantis 12,000 years ago.

Dr. Jesenský, mentioned here in his work, claims that our civilization grew on the ruins and ashes of the previous Atlantean civilization. These supposedly left gigantic artifacts in the form of a stone sanctuary in Marcahuasi, pyramids in Giza and other places, stone balls in Mexico and Slovakia, Observatory in Stonehenge, mounds in Lesser Poland, etc., which sleep on the eyelids with archaeologists and seekers of the Unknown.

This civilization burned in the heat of a thermonuclear war that transformed into deserts the vibrant areas of Africa, the Americas, Asia and Australia. Perhaps it was people who were attacked by aliens from outer space, which the Earth gave to settle and colonize. We probably will not find out this way. The areas left behind are still relatively untouched by the war disaster and Cappadocia is one of them. It is here that the people who survived the Great Conflict hid, and here they survived the worst period of climate change after the end of the IV Glaciation (it is called the North Polish Glaciation). It is possible that these cities were built even earlier - 1.8 million years ago, when the first or second glacial period began (in Poland: Przasnysz and Południowopolskie [Krakowskie]) - the worst of all - which embraced half of Europe and sharpened the climate on the Mediterranean Sea. Either way, the construction of these cities is closely related to climate change on Earth and only a nuclear war would cause cataclysmic climate changes comparable only to the effects of an asteroid.

It is interesting because looking at all these buildings, we were reminded of the report by Dr. Ludmila Szaposznikowa about the mysterious land of Shambhala - Szanszun or the Land of the Bus (which James Hilton named Shangri-la in his novel "Lost horizon"), in which the wise lived in the hollow rock spiers above the sacred Nam-co-to-rin lake. Almost everything agreed, there was not just the holy lake Nam-che.[14]

One more thing, in our opinion, is a mysterious fact which our guide told us, namely: prolonged staying in the corridors of underground cities causes irradiation of the human body with low doses of ionizing radiation, which was discovered during periodic medical examinations in Ankara. It is natural - tuffs, obsidianates and other volcanic rocks always radiate more strongly than the normal radioactive background of the Earth, which can be seen in the High Tatras, whose granites radiate 21-30 μSv / h and more. In the case of Cappadocia, the doses are stronger and longer exposure to them poses a threat to human and animal health. Could these be tangible traces of the Great Conflict?

Anyway, it turns out that these underground cities were just shelters from radioactive rainfall, which from time to time haunted this place. After all, from Cappadocia, there are deserts of Saudi Arabia, Rajasthan (Thar) and Sahara, which - if we have to believe the old texts of the holy Hindu books and legends of peoples of Africa - they were the targets of the strikes of weapons of mass destruction in the Great Conflict of Astronauts before 120 centuries. So, the underground buildings of Cappadocia are another link in the long chain of evidence that we are the first on this planet.

* * *

---

[14] See also FA Ossendowski - "Through the Land of People, Animals and Gods", Poznań 1930.

Please note the last paragraph. These words were written by a professional historian.

CHAPTER III

Spherical puzzles and other

For many years, scholars have been trying to guess the origin and destiny of objects and biological phenomena that have been discovered as being "out of place" in the world. And again in 2005, a discussion about the origin of metal balls from South Africa started. Their diameters are 2.5 - 10 cm and look as if they were made by a man. These only could have arisen in Precambrian, much earlier than Homo sapiens appeared on Earth because they are counting from 2.8 to 3 billion years! The discovery of this and other artifacts is treated by the article by Marek Sokołow published in the pages of the Russian journal NLO nr 16/2005.

Model of Death Star from 3 billion years ago?

The existence of these metallic balls became apparent already in 1977, when it was necessary to save ancient petroglyphs on a pirofillite (a wonderful stone - as they are called there - car notes), and whose huge blocks began to cut into smaller ones. Everyone was amazed at the perfect shape of this ball, as well as the incision made only in the middle.

In the following decades, South African miners found at least 200 such balls. Some of them have up to three parallel grooves on their "equator". Some of the balls were halved and it turned out that they are covered with a 6-millimeter thick armor. In the middle was a spongy material that turned into dust in contact with air and it resembled charcoal in its appearance. Other balls turned out to be full and made of bluish metal with white spots.

Hard ball to crack

What was this metal? The analysis showed that it was a steel and nickel alloy but in Nature, something like that never occurs. From the point of view of Rolf Marx - the custodian of the South African Museum in Klerksdorf - these balls are the truest riddle. Those that are in his museum themselves vibrate in some strange rhythm, although they are cut off from any sources of energy. Maybe some mysterious energy is hidden in the interior of the balls, which has not yet worn out despite the passage of 3 billion years?

One time, a well-known sorcerer was brought to the museum, who said that the ball came to us from space and has a magical power.

In 1984, in response to an inquiry by one of the editors, R. Marx wrote that there are virtually no scientific works on these spheres, but such spheroids can be found in the pyrophethalmus excavated near the town of Ottosdal in West Transvaal. Pirofillit with the chemical formula $Al2$ $[Si4O10]$ (OH) 2 is a soft mineral. The balls are characterized by high hardness and it is difficult to scratch them with a steel blade.

The second puzzle of beads - cut resistance. You have to try to find some explanation for the creation of these crèches around the spheroid, especially those three that are parallel to each other. If we do not find a scientific explanation for this problem, then we will have to admit to something mystical, namely, that these balls, which today are in deep layers of rock, have done billions of rational beings before billions of years.

Fifteen years ago, these spheres attracted the attention of John Hund from South-African St. Petersburg. He went to the mine and found such a ball, exactly as it is in the museum in Klerksdorf. Once, when he was sitting at a table in a restaurant, he took out a ball and put it on the table and noticed that the ball was very stable. Hund decided to send the artifact to the US Space Research Institute at Caltech, California. It turned out that the ball is perfectly balanced. Balance accuracy reaches 1 / 100,000 inches! A NASA scholar admits that there is no technology in them that would make something like this possible on Earth. Something like that can be done but only in the conditions of zero gravity. Under conditions of weightlessness. Only in space!

Directed panspermia and time machine

The Dutch scientist BV Lourker assumes that some three billion years ago these balls were brought by extraterrestrials - aliens, newcomers, aliens. For what? These were containers of single-celled organisms, the most primitive forms of life. Now let's look at it this way - if they had the technology then, what have they got today? After two - three billion years of its existence?!

Well, if it was not them, maybe it was done by some ancient race of Earthlings, who lived long before us on this planet, and which either died or left the Earth. (As in R. Cook's novel "The Abduction", in which so-called People of the First Generation occur, who have hidden themselves under the surface of the Earth - attention of the RKL)

Perhaps these balls also formed themselves as a result of some natural process like spheroids on Mars. But why is this museum ball is so stable - due to the action of magnetic fields?

Some scholars who succeeded in studying these small balls came to the conclusion that these spheroids were made artificially and not due to natural processes. And maybe they do not come from the Past, but vice versa - from the Future! The crew either died or somehow managed to get back to their time, leaving the details of their vehicle in the past, and maybe even the whole vehicle?

A mundane explanation

There are also more mundane versions - in the literal sense of the word. Since the Earth has undergone many terrible cataclysms during its history, its geological layers could have been mixed up many times and changed places. Therefore, some modern objects could get to the oldest rocks on Earth, due to a pure case. Researcher Paul Heinrich wrote five years ago that an unhealthy noise rose around this problem, caused mainly by a variety of sensationalists focused on the tabloid "Weekly World News." Errors and inaccuracies were even found in the book by M. Cremo and R. Thompson entitled "Forbidden archeology" and in the NBC program entitled "Mystical origin of man." It is not true that there is no scientific literature on the subject of African spheroids. She exists. It is also clear from it that pyrophyllite is a metamorphic rock, not a sedimentary rock. This metamorphism manifests itself at moderate temperatures at a depth of several kilometers. With the help of geologists from South Africa, Heinrich came to the conclusion that the mysterious balls consist of pyrite - iron sulfide - $FeS2$ and goethite (goethite) - iron oxide - FeO (OH). In the process of transformation of clay or volcanic ash into pyrophyllites, pyrite spheroids formed in the rock mass. And from the nested pyrite deposits, which were pushed out into the oxidation zone, goeth was formed. Thus, these are not any concretions - as was written in a whole range of studies.

Why are these balls so hard? This is because goitic balls can absorb other hydroxides, which, when bound to them, increase the hardness of these beads and the English decided to do business based on these balls. They produce balls with a diameter of 7-15 cm from pyrite and sell them to singularities and lovers of exoticism for a price of 100 USD per item.

As for the spherical shape of these spheroids, Nature is able to shape its creations arbitrarily and artistically. It's enough to mention just the huge stone balls of Costa Rica. However, the largest stone ball is in a quarry near Klokočov in Kysuce in Slovakia which measures 305 cm in diameter and consists of brown magurian sandstone. It must be added that the full-metal balls sometimes fell from the sky in different countries of the world. Three such shiny, polished spheroids with a diameter of 35 cm each were found in the Australian desert in 1963. Speaking in Parliament, the Australian Defense Minister Allen Feihell admitted that all attempts to open these bullets came to naught. This was described by Lucjan Znicz-Sawicki in his work "Guests from Space - NOL" - note: RKL. There are rumors that these bullets have been handed over to the USAF and that hearing about them has been lost. Small, colorful spheres were found on the territory of France and in such an amount that it looked like it had fallen with rain. As you can see - Nature loves to play balls with us!

However, this is just the beginning of the puzzle. In 1972, chemist analyst Boughzighe working at one of the nuclear fuel processing plants in France, pointed out the uranium ratio U-235 to U-238, unusual for uranium ore. This unusual anomaly is the subject of an article by Irina Stiekałowa published in "NLO" No. 3/423 dated January 16, 2006.

Physicists have let their imagination run

Of course, a few hypotheses were immediately made about this strange anomaly of an unusually large amount of uranium-235 in this ore. The first of these was the hypothesis that the ore was contaminated with overworked nuclear fuel. However, accurate radiation measurements showed that such an explanation does not explain anything. It turned out that uranium ore with an unexpectedly low percentage of uranium-235 was extracted in Gabon - one of the equatorial countries of Africa. Why this ore is depleted of this isotope remains a mystery.

The appearance of this anomaly in Africa could be explained by assuming that the field of "normal" uranium ores was irradiated with neutrons. Such a paradoxical hypothesis was put forward by the Russian physicist NA Vlasov. In the scientific community, it was sensational. To the physicists-atomists, dozens of uranium mines were known, and in them the amount of uranium-235 is always at a constant level - about 0.7%. In the anomaly discovered in 1972 in the Oklo mine, the content of the uranium-235 isotope turned out to be only 0.44%.

Physicists have let their imagination run wild. There were hypotheses saying that the deposit was contaminated with overworked nuclear fuel from an extraterrestrial spacecraft, or that there was a "graveyard" of radioactive waste from ancient civilization.

"There was a nuclear explosion here," some said, "and it happened millions of years ago."

"No, this is the job of a natural nuclear reactor," said the others.

It just perfectly explains the partial burnout of nuclear fuel. Exactly the same process takes place in normal reactors of nuclear power plants!

Vlasov's hypothesis

Soon, the most interesting was the third hypothesis, which assumed the existence of antimatter in Nature. It is known that physicists-theorists assume and allow existence in the vastness of the universe of antimatter. At the meeting of matter with antimatter, annihilation should occur - an explosion that will result in the passage of matter into energy, according to the Einsteinian formula for the equivalence of matter and energy: $E = mc^2$. This process is characterized by a whole series of nuclear transformations and the emergence of colossal amounts of energy, in accordance with the above-mentioned formula.

However, it is small but in conditions prevailing on Earth, such an event is almost unbelievable. Nevertheless, hypotheses appear from time to time, whose authors allow a slight probability of an event of falling pieces of antimatter from space, such as an anti-meteorite, onto Earth. Such a hypothesis has already appeared in connection with the decline to the so-called Tunguska Meteorite.

(=> Antology "Siberian Bolide", Jordanów 2001 - Internet - http://www.sm.fki.pl/Lesniakiewicz/Lesniakiewicz.php?nr=Bolid_Syb eryjski - attention from the crowd).

The annihilation collision of an antimatter meteorite falling into the Earth was characterized by a series of nuclear transformations that caused a neutron flux.

According to the assessment of NA Vlasov, the neutron flux of incredible intensity could be caused by the fall of anti-meteorite of 1 ton on the Earth. (Annihilation of only 1000 kg of matter would release energy equal to approximately 90×10^{18} J, or about 21.5 Gt of TNT - many times more than the entire nuclear arsenal of the Earth, whose power according to calculations by SIPRI in the 1980s only 9,4 Gt TNT - crowd note). The anomalous content of the U-235 isotope can be explained by an explosion of this amount of antimatter. (The energy of the Tunguska Meteorite explosion - if it was a meteorite, and not something else - amounted to only 13 to 130 Mt TNT according to various estimates. Therefore, it was a meager fraction of the energy of the explosion - 1 ton of antimatter - crowd note)

However, more accurate studies have shown that this unusual ore was characterized by a strange chemical composition. Analysis of the uranium decay products present in it revealed that around 2 billion years ago, nuclear chain reactions occurred around the deposit.

Nuclear physics is a relatively young field of knowledge, and the first nuclear reactor was constructed in 1942. Nevertheless, nuclear reactions took place on our planet already 2 billion years ago - in Proterozoic. (This is the youngest period of Precambrian, which lasted from 1,900 - 600 million years ago, 2 billion years indicates that the deposit was created in the earlier pre-Cambrian period, and precisely at the end of the Archaic period [2.7 - 1.9 billion years ago] - RKL). So far, science knows 17 former nuclear reactors located in the south-east of Gabon. All of these reactors were found in the area of uranium ore occurrences in Oklo and Bangombe. Nine of the 17 reactors were discovered in the excavations of old uranium ore mines in one of the countries of Equatorial Africa.

Dating of natural nuclear reactors is possible due to the fact that a long time ago there was more U-235 isotope on Earth than today. The chain reaction takes place in the moment when the amount of this isotope is greater than 3%. It is in such an enriched ore that the optimal neutron uptake reaction is sustained. The operating time of the Gabonese reactors is millions of years. Nowadays, the possibility of such reactors is small, because the concentration of 235 U isotope in uranium ores is already small, and there would be no spontaneous chain reaction.

Natural nuclear reactors represent immense value for science. These are unique objects, not found anywhere in the world! They allow us to get to know the many secrets of the history of our planet and help us in the operation of our nuclear reactors nowadays. In addition, to examine these reactors will help us develop procedures and technologies for the utilization of nuclear radioactive waste.

So much Irina Stiekałowa.

For us, this case began with the reading of Jerzy Edigey's novel (1912-1983) entitled "The guardian of the pyramid (second edition, Warsaw 1990), which - as we expected - would be similar to the famous "Arrows from Elam". It turned out that the thing was about building Egyptian pyramids and protecting them against robbers using the "land of death" - radioactive uranium ores brought to the Land of the Pharaohs from ... Gabon! After I found out about it, God and the topic was exotic, and even far from Poland. It was only after 1978, when Polish and foreign production began to appear on our book market on subjects forbidden by scientific censorship (and it was even more severe than social or political!), that we learned a little bit more about it. But the thesis about the natural origin of these nuclear reactors was still being promoted. Well, because everything is clear - reactors are natural and there is nothing unusual in them. But are you sure? Dr. Miloš Jesenský writes about it in one of his books:

This is still not all God. We have premises to believe that the Earth's nuclear energy is much, much older. A near-detective investigation began in May 1972, when routine analysis of uranium ore was made at the enrichment plant in Pierrelatte, France. The results of this analysis stunned and amazed physicists and researchers at these plants. It turned out that in the unexpected analysis, the percentage of U-235 isotope is only 0.717%, not 0.720%, as it is everywhere on Earth, the Moon and meteorites.

So, where did the remaining 0.003% of U-235 die? The trail led to Gabon, to the uranium ore mines in Oklo, from where the ore samples were taken for research in French laboratories.

The ore analysis was carried out immediately on site. The surprises of professionals have grown to morbid proportions, because research has shown that 700 tons of uranium, which was excavated there in 1970-1972, lack 200 kg of U-235 radioisotope. Little? Such amount will be enough to construct several dozen Hiroshima atomic bombs (So about 20 kt TNT - attention RKL)

It seems that no one in the recent past from Gabon has stolen such a quantity of U-235, and thus it would seem that in uranium ore in Oklo there is less uranium than it should be there! And that means only one thing - someone once separated uranium-235 from natural ore!

Although the experts hurried to explain the "natural nuclear reactor at Oklo" immediately, the "natural nuclear reactor at Oklo" hypothesis assumed that in the vascular uranium ore [uranophane, torbenite], earthquakes formed deep fissures, which were then flooded by rainwater, stopping neutrons. This was the beginning of the U-235 uranium chain decay reaction and the release of large amounts of thermal energy, which then evaporated the water and the reaction stopped.

Repeated from several thousand rainy seasons, the cycle of the "natural reactor" activity could have caused uranium ore depletion in U-235 by 0.003%. So far, no one has verified this hypothesis and is based mainly on the desire of scholars. It is hard to believe that to this day there has not been spontaneous nuclear explosion and contamination of surface and groundwater with uranium kernel decay products. Something like this has not been found on Earth, although we know a few places where such a process would be possible - editor. RKL) remains a fact that someone or something in the cavernous past separated the U-235 isotope from the rest of the ore exactly as if he wanted to use it for energy purposes.

Of course, scholars have tried to explain this phenomenon by the fact that uranium ores in Oklo consisted in mass with 17% U-235 and 83% U-238. These isotopes have different half-term time - T1 / 2 - and before 2 billion years, this uranium ore contained only 3% U-235. Uranium could be washed away by water in the delta of the ancient river, where today lies Oklo and there settled U-235 enriching it. When the amount of U-235 increased to critical mass, a chain reaction occurred.

However, there is no such reactor, the fact remains that we know little about nuclear power in Antiquity. (M. Jesenský - "Gods of Atomic Wars", Krásno by Kysucou 1998-2001, Internet http://www.sm.fki.pl/Lesniakiewicz/Lesniakiewicz.php?nr=Bogowie_ Atomowych_Wojen)

Therefore, nothing is clear. Nobody answered the basic questons that should be:

1. Why did these reactors originate in Gabon and not somewhere else? After all, since these are natural formations, they should be created where uranium ores are found. Meanwhile, they are found only in Gabon.

2. In which Central African country are there still such reactors? The map shows that such a country may be Niger (or more precisely the village of Alit, in the Saharan mountains of Aïr), where there are huge deposits of ores of this element. But are you sure? I have not come across any information that such natural reactors are also found there.

3. What is the strangeness of the chemical composition of ores in Oklo and Bangombe? If they contain only products of natural decay of uranium and thorium, this is not surprising because these radioactive ranks are predictable and computable. However, if there are any other elements from outside these ranks, archaeologists should also take care of it.

In one of his articles, Robert Leśniakiewicz suggested that the best sign of staying in the Solar System of a higher Scientific and Technical Civilization would be to find isotopes of elements in Nature that do not exist and were synthesized in nuclear reactors. Could this be the accident I described in the case of "natural" nuclear reactors in Oklo?

If so, would it be a proof that these reactors are not as "natural" as scientists say about them, and perhaps they are an atomic "graveyard" of nuclear waste of the Atlantean civilization and the creators of the Lunar Well?

But what about reactors that are thousands of years old? Walentin Psałomszczikow writes about iron figurines that count half a billion! Recently, an interesting letter came to the editor of the weekly "NLO". Its author was Innokientij Michajłowicz Połoskow - the main geologist of the West-Lipoviensko Mine located near the village of Pobugskoje in the Gołowaniewski District of Kirovograd Oblast. In earthworks at a depth of 30-40 m from the earth's surface, he found dozens of figures from 1 to 40 cm in size depicting images of various animals and birds, including extinct ones, as well as figures of people.

Probable age...

...these figures are from 100 to 600 million years![15] They were found in layers belonging to the Proterozoic, which cut into the Tertiary rocks, which are about 100 million years old. These figures were made from minerals containing silicon with admixtures of magnetite and chromium compounds and hydroxides. These figures were found during the two years of this work.

In 2002, scientists from the Kiev University conducted a study of magnetism, as a result of which it was established that the strongest he was in the place where these figures were. Until now, the amount of iron in the examined ore deposits did not exceed 30%, while in the places of hematite nests, where its amount was 40-47%, the magnetic field intensity was much weaker.

In order to explain the nature of his discovery, Poloskov put forward three hypotheses:

1. A dozen or so millions of years ago, landings of Przybyszów from space have been planted on Earth, who for unknown reasons have left metallic figurines of animals and people from their planet. Over time, the action of water, microorganisms and other factors caused the ingrowth of other minerals, especially quartz with inclusions of other metals. Incidentally, in the magazine "Znanie-strength" from 30 years ago, there was a photo of a typical rifle bullet found in a block of coal. The missile hit a tree that after several thousand years was petrified and the metal of the projectile was replaced by silica.

2. According to the hypothesis of the employees of the Moscow Institute of Medical and Biological Problems - A. Belowa and W.

---

[15] The Tertiary lasted from 64.8 million years to 1.4 million years ago.

Vasiliev, man appeared on Earth about 500 million years ago, and the world of animals proper for this period was created.

3. Figures of people, it's just nature's play.

But immediately, the following questions arise: why are these figurines in a strictly defined place size 30 x 50 m, although these mineralization processes also occur on other sections? On the figurines of birds, the proportion of the torso to the wings and beak is clearly visible, and the eyes are visible in the form of black inclusions. The same is true for dog figures - eyes and ears are in the right places.

Return of artistic strength?

The answer of the Polovtsian letter was accompanied by the response of a research worker from the Crimean Division of the Ukrainian Scientific and Research Institute of Geological Research, Dr. W. Szirokow, Ph.D., geological and mineralogical physicist. The author of the answer supports the hypothesis about the artificial origin of these figures.

Indeed, it could not be just a fool of Nature because some figures look like they were cast in forms, which can be seen even in the pictures. Why had Nature decided to deploy these figures in such a small space?

However, the Poloskov hypothesis regarding paleo-contact has some weakness, namely: why the aliens would be doing figures of Earth animals and birds - including penguins, bears, dogs, cats or seals and humans. All of them have clearly exposed eyes and above all, this stone zoological garden resembles quite accurately some decorative elements or decorative figurines, which someone smoked from clay as toys for their children. Of course, where did they come from, half a billion years ago? Could geologists have been so mistaken in assessing the age of the find? A Western specialist invented an original hypothesis that some of our ancestors-jokers made these figurines and buried them to ridicule scholars. Just the remaining question, how were they able to dig a hole to a depth of 40 meters without shoring and securing it?

The second skeptical hypothesis is analogous: figurines were made and buried by modern counterfeiters.

Stones from Ica

In fact, the story is repeated with American stones that were discovered in the area of the Peruvian city of Ica. On their black surface, with some very hard tool unknown to us, impossible from our point of view and our historical knowledge engraved drawings, such as people hunting for dinosaurs, flying on fiery dragons or winged reptiles, elephants, camels, kangaroos - unknown on the American continent, and also animals extinct hundreds of thousands and millions of years ago. There are also maps of continents, which are very different from those known to us today. On the other hand, some human-like beings perform heart transplant surgery. The dating of these stones is complicated: numbers in the press have been published for tens of thousands to several million years. These stones were collected and placed in a museum of about 15,000 but according to expert, estimates over 50,000 are in private collections.

Again, there are hypotheses about their falsification that the locals are ungraded Indians who are rooting these pictures on stones, and then they bury them to sell to many tourists. Only from these dark Indians, such historical knowledge - they did not see dinosaurs or elephants and camels in their lives!

Well-known finds for a long time

We often wrote about similar finds of various strange artifacts on the pages of "NLO". It should be added here that even the well-known head of "Kosmopoisku" W. Czernobrow unearthed something similar on our territory.

You can also mention old, previously known findings. In the middle of the 19th century, human skeletons were found in Switzerland and California in the geological layers created 35-50 million years ago. In 1910, metal pipes were found in the area of the French town of Saint-Jean-de-Luves in the chalk layer from 50-140 million years ago. In the layer, about 300 million years ago, a metal disc was found. American geologists found a shoe imprint with a characteristic protector, in a rock of 300 million years old. There are many similar finds, only geologists and archaeologists do not want to admit that they exist and, of course, talk about fakes.

There is still one moment in our history that immediately strikes those who have perfectly mastered school textbooks: figurines, whose age is several hundred million years, imagine not only people, but also birds and other animals with whom they seem to share this the same epoch. Yet, according to the book's knowledge, the first was the age of dinosaurs, which later - for reasons not fully understood - were replaced by mammals. Was it the reverse and warm-blooded mammals were the ancestors of dinosaurs that were born as an evolutionary counterproposition? And the explanation of this phenomenon is - so far - only one thing: after some global catastrophe, there existed a few hundred million years ago civilization, which was completely lost and the development of the Earth's flora and fauna began almost from scratch. In life, some specialists will say that they are falsifications or games of the forces of Nature, while in the meantime, there are hundreds of thousands of such artifacts that are just waiting for their explorers.

We would go over this information if it were not for a significant fact. He brands it with the name of the PhD of mathematics and physics. Valentyn Psalomyshikov is one of the two authors writing for "NLO" and giving their scientific degrees. He is one of those scholars who is not afraid of losing his reputation in the scientific world, and that's what counts in such cases.

CHAPTER IV

Traces in the micro world

In his work, Fri The Atomic War of the Gods (Lublin 1979), Aleksander Mora, claims that several thousand years ago, a magnificent civilization flourished on Earth, which killed itself in a titanic conflict with the use of all known and unknown weapons of mass destruction.–I It was destroyed in a total war with cosmic colonizers and as a result of civilization regress - she went back to the Stone Age. Similar views were also given by Dr. Jesensky in the work "Gods of Atomic Wars" (Ústi by Labem 1998). Its enunciations of those contained indicate that the civilization of Atlantis or the preceding civilization of the Atlantic reached our nearest planets and colonized them. What is worse - it led BMR into space, which could lead to inactive and unused units of these cosmic WMDs still circling the Earth, the Moon and other planets of the Solar System, threatening to destroy everything that lives on it. The effects of dropping warheads A and C have already been described, which can be found in Robert Leśniakiewicz's paper "Project Tatry" (Krakow 2002), while in the pages of "Peripheral Vision", he described the operation of the weapon B. Now I would like to focus on this topic because recently, we have obtained evidence that biological weapons were created by previous civilizations, and that as a result of The Great Conflict this weapon has escaped from human control.

Aleksander Mora wrote about it:

The ancient Hindu text, "Samara Sutradhara" clearly speaks of the use of biological weapons in the distant past - B. A specific named Samhara was used as a disease-causing agent among the enemy's soldiers, while the other - Moha - caused numbness and paralysis.

"Fengshen-yen-i" mentions military operations in B carried out in China and again, in these texts, we find descriptions remarkably similar to Hindu ones. At this kind of occasion, the question arises: is it not possible that some contemporary humanity-afflicting diseases have in the past been induced in an artificial way? There are many diseases that only people undergo but they do not bother animals. Could they not have arisen as a result of an ancient, destructive bacteriological war, the scope of which escaped the fighting parties out of control?

Well-known biologist A. Firsow draws attention to the fact that viruses, currently regarded as representing an intermediate stage between the living world and the deceased or the organic and inorganic world, behaving in an inactive state like crystalline substances. I in the active state reproducing and showing targeted actions do not have to arise at the dawn of life on Earth. The more so because they show a high degree of specificity towards the host, which could indicate their relatively recent origin.

Another mysterious matter, closely related to the atomic tragedy in prehistoric times, are paintings on the walls of caves and on rocks scattered all over the globe. These paintings, depicting the forms of the so-called Kosmites have become world renown in recent years, and their special popularity dates back to the publication of the book Erich von Däniken entitled "God's Chariots". One of the best known is the outline of a giant figure engraved in the rock. It was discovered by Henri Labote on the Tassili plateau in the Sahara Desert and named by him the Great Martian God. This sketch surprisingly resembles a figure dressed in a space cosmonaut suit.

There are more similar drawings in the area of the plateau: one of them depicts a group of four figures dressed in clothes resembling cosmonaut suits, whose heads are covered with spherical helmets. Drawings of this kind have been discovered on various continents. There is, for example, a rock drawing south of the town of Fergana in Uzbekistan, showing a figure whose head is surrounded by a ring with rays radiating from it. This ring most probably represents the helmet worn by divers equipped with antennas.

Almost identical figures depict drawings discovered in Val Camonica in Italy. The silhouettes of the Kosmites were also found on flat rock walls in Australia.

Significant resemblance to the figures in the drawings show Japanese Dogu statuettes from the Jomon period (Dzomon). They aroused great interest because they present people in some kind of protective attire and helmets with strange glasses. The Japanese expert, Isao Washio, describes Dogu's outfit like this: the gloves are attached to the forearm with a binding, and the glasses can be closed and opened. On the sides of the figures, there are levers probably designed to adjust their position, while the "crown" placed on the helmet acts as an antenna.

The devices outside of the clothes are not decorative elements, but are devices allowing to control the pressure in the suits in an automatic way. All these drawings and figures are now associated with the visitors from space and the view that our planet in the distant past visited guests - astronauts from other star systems. In fact, they can represent celestial gods - the more so because the drawings of the Kosmites are often accompanied by flying disks, spherical vehicles and other flying devices. It seems that there is another, no less likely explanation, based on a different interpretation of the found drawings.

Perhaps they simply represent people in garments that protect them from radioactive contamination. After all, most of the drawings of the Kosmites were discovered in the area of today's deserts. Previously, however, it was suggested that the first reason for the creation of these deserts could be the use of nuclear weapons in these areas, which is why even thousands of years later these regions indicated a high degree of radioactive contamination.

The drawings could then be made later by the descendants of the inhabitants of these regions who managed to survive the cataclysm. They saw those arriving (either from underground shelters or from areas unaffected by radioactive fall-out) of people in flying machines who, protected by protective clothing, appeared to inspect the areas, investigate the extent of their contamination and assess the extent of damage.

Undoubtedly, some of these drawings may depict Przybyszów from space. However, it does not seem right to overlook the possibility that protective clothes were worn by the inhabitants of the Earth as a shield against radioactive contamination. For if the aliens had to be so carefully protected from our terrestrial environment by means of such insulating cosmic suits, it would mean that they could not breathe the Earth's atmosphere or bear the temperature prevailing on the surface of the Earth. And such a picture of cosmic gods - as Richard E. Mooney writes - would be completely in contradiction with the one we have developed on the basis of myths and legends.

The mythical gods of Egypt, Greece, India as well as the Maya, Inca and Aztecs- never were they described having protective clothing. That is why they were either Earthlings or visitors from the

Cosmos, whose physiology was similar to Earth's in such a degree that they did not need protective suits. Of course, those who would only come for a short stay on Earth, could wear clothes that would protect them against different solar radiation or unknown, and possibly dangerous to them earthly microorganisms. However, taking into account the arguments "for" and "against" and considering the distribution of the main paintings and material evidence of a nuclear disaster, it seems that the most rational explanation for the functions of these "space" suits is protection against radial contamination.

Or maybe it would rather be a biological contamination? Protective suits against microbes are as tight as cosmic or deep suits, and therefore are similar to them and equally functional. Beings who were afraid of terrestrial bacteria could not be aliens, but humans. The fact that these people wore suits is proof of this. Earth's bacterial flora may be particularly dangerous for beings that have similar or the same metabolism and cell structure similar to bacteria, and therefore, bacteria are unable to infect a foreign organism, because they simply could not multiply in it, and as a result would quickly die.

Anyway, these malicious microbes were created as a result of genetic manipulation, which may have been the case, according to Al. Mory:

The earth was badly damaged. The main centers of civilization did not exist. Centers of disposition, some land and all cities either disappeared under the waves of the oceans or turned into ash scraps under the influence of laser and nuclear weapons. The rest of the destruction was caused by volcanic eruptions, shaking the earth's crust, whose balance was so recklessly violated.

Despite the frighteningly sad balance, new life began to be organized. The survivors of equipment and devices were collected. Technical crews penetrated the Earth's surface, looking for those who managed to survive the cataclysm, as well as raw materials, propulsion, medicines and food. Construction of new cities was undertaken, such as Tiahuanaco and estates like Sacsayhuaman. Life began to enter into more stable modes, despite the fact that not many societies had to deal with a whole range of serious problems.

First of all, it was not uniform. His was composed of people who were brought up on different planets. Some of them from the beginning could not adapt to breathe in the Earth's atmosphere, probably they had to use assistive and protective devices. For others, the solar radiation on Earth was too strong. For others - too weak.

All these difficulties were to be remedied as soon as possible, if the very few people were to protect themselves from total extermination and extinction of themselves and the achievements of a civilization lasting thousands of years. Among the tropical forests, forests and savannas of the Earth existed primitive human tribes at a low level of development, whose way of life did not differ from the animal. So, it was decided on a bold biological experiment, the aim of which was to perform a genetic surgery on several selected tribes, allowing for a significant acceleration of their development. Obtained in this process individuals were to be used in the future as a cheap, unskilled workforce, understanding and performing properly and in a disciplined manner set by the god-creators of the task.

The experiment succeeded well, and its effects - it seems - exceeded all expectations. Intellectual development of the tribes subjected to genetic surgery in a limited area, followed by careful care and education proceeded very quickly. At the same time, he led to the creation of an unforeseen by-product - he solved the problem with which the gods have so far struggled, because he provided a host of extremely beautiful women. No wonder then that the younger gods immediately proceeded to improve the results of the experiment, forgetting that although they belong to the same genre roughly, they represent different paths of genetic development, which particularly concerned those who were descendants of generations long-living to others, Earth, planets.

The Biblical "Book of Genesis" gives: And when people began to multiply on Earth, their daughters were born. The sons of God, seeing that the daughters of a man are beautiful, understood them as wives, all they liked. In those days, there were giants on Earth because when the Sons of God approached men's daughters, they gave birth to them. They were, therefore, these powerful men who had fame in those old times.

The mutants born of these relationships could not always be the object of the boasting of the gods. Some of them, being completely degenerate forms, had to be liquidated. However, most of the descendants represented a greatly increased intellectual capacity, some had such a high intellectual level that they were easily mixed up with the community of younger gods.

A successful experiment caused a new, rapid development of civilization on Earth. The gods already had a handful of manpower. Selected from among the masses, people obtained technical and even scientific qualifications; they did engineering work, they were pilots, soldiers, doctors, and finally they achieved the status of god. The old gods were dying out, the young have become more and more infused into the vibrant society of intelligent Earthlings.

Since the people of that era were able to manipulate the genes to grow Homo Sapiens, breeding a malicious strain of bacteria or a virus would be an easy task for them! And it was!

It has always been a mystery how it could be proven. And such evidence was found in books by Richard Preston - "Zone of contamination" (Warsaw 1996); Peter Radetsky - "Invisible invaders" (Warsaw 1998), and above all in the work of Christopher Wills "Yellow Fever, the Black Goddess" (Poznań 2001).

Well, one of the most lethal diseases are bacterial infections in the genus of Asian cholera - Vibrio cholerae, or bubonic plague - Yersinia pestis. Epidemics and pandemics of these diseases have decimated the entire world or even ruined its vast areas, because plague is also a disease that attacks animals and kills them - and above all those that accompany humans: dogs, cats, mice, rats, horses, cows and pigs.

There is no vaccine for cholera today - you can only treat it with antibiotics. Scholars have not answered the basic question so far - since when does cholera kill people? Because only one thing is certain - it was not always harmful to people. It is known that relatively recently, it broke the species barrier but when was it? Nobody knows that.

The plague is a bit different. Its lethality is the result of not being added to a gene that causes "malice", but genetically deprived of its ability to move in a liquid environment, also lost many of the biochemical possibilities of changing its host to another (in other words, "assigned" to specific animal hosts), which also means that it cannot survive in the soil; also eliminated her Krebs cycle, which decides the receipt of many components that arise during breathing. This means that it must receive them from the host. And further - Y. pestis cannot produce the hyaluronidase protein itself, which would allow it to penetrate to the host cells. And this has very serious consequences, because it increases the murderous abilities of these bacteria. This has been proven very simply - the genes responsible for the production of hyaluronidase and the Krebs cycle have been destroyed in bacteria related to bacteria Y. pseudotuberculosis - which caused them to become as lethal as Y. pestis.

The virulence of Y. pseudotuberculosis administered orally increased by 1,000 times and injected subcutaneously - as many as 10,000 times! Research on these bacteria was carried out not only on mice and laboratory rats.

The conclusion drawn by the scholars was one - the emergence of two mutations of one bacterium at a time is highly unlikely, and therefore they were not the work of Nature. The bacterium Y. pestis has been deliberately genetically damaged to raise its malice. And what follows - all the above genetic conditions make it an ideal biological weapon, which was created only to destroy large human populations, to kill everything that lives and then self-liquidate for lack of hosts. For what? - It is known: so that aggressors using this BMR can quickly master a given area. So it was definitely the first-strike offensive weapon or the ideal weapon for subversive action!

There is one more case related to Y. pestis - its main vector is fleas. This bacterium can modify the organism of this insect so that the flea infects as many living creatures from its surroundings as possible. It completely clogs her intestines leading to rapid dehydration, which forces the flea to eat as intensely as possible - jumping from the host to the host - and infecting many of them before they die. Chicks, mice and cats, in turn, infect people who are the target of a biological attack.

Note that both of these God's deeds have a very short incubation period: cholera 2-3 days, and plague 2-5 days. As you can see, these two bacteria are ideally suited for biological warfare through so-called biological diversion - dissemination of diseases using subversive and reconnaissance groups, subversive bombardments, missile fire, etc. After aeration of a given area epidemic breaks out, and then - when death collects its harvest - the aggressor's army enters the area.

Researchers concluded that all manipulation causing the non-malignant strains of bacteria to grow is not so complicated to perform, but extremely effective in use. The same can happen with viruses that are even more adapted to the conditions of a bacteriological war and are almost perfect weapons B.

What the hell is this bacteria? It has an additional gene that is responsible for its ability to synthesize a toxin. Christopher Wills simply states that this fragment of DNA has been in the helium of V. cholerae for a long time, adding that it has acquired it from another bacterium. But again - this possibility is rather slim, so it seems that someone "helped" V. cholerae this transposon can be purchased.

Looking at this nasty set of genes - he writes - almost has the impression that this transposon has been deliberately designed to give Vibrio the properties necessary to wreak havoc on human guts.

It will not be mentioned here that there are also animal diseases that are caused by bacteria from the Vibrio and Pasteurella strains. They are usually fatal with a 99% effect.

Other nasty pathogens are typhoid fever - Rickettsia prowazekii and typhoid fever - Salmonella typhi. The latter is particularly nasty, because S. typhi can stick to the walls of cells lining the intestines of the victim, penetrate inside the body and finally is able to attack those cells that could neutralize them, which is the elimination of the immune barrier - as in the case of virus HIV! Again, there are two types of S. typhi strains - African and those that inhabit Africa and the rest of the world. The latter appeared on Earth in the period from 2,000,000 to 200,000 years ago! This world strain is able to infect its victims and enter the gall bladder, and from there expelled from the host organism and lead to contamination of many people. African tribes cannot do it. So the next intentional mutation? Again, studies of the DNA of Salmonella and the related Shigella flexnen bacteria (causing dysentery) confirmed this thesis. The DNA of both bacteria contained additional genes responsible for their murderous properties. So far, scientists do not know how these mutations came about but no one thought about their artificial origin. It was not the evolution that gave these microorganisms their murderous properties, and the purposeful robot.

The Escherichia coli bacterium is somewhat less dangerous, which is responsible for some food infections. Scholars believe that it was the "genetic source material" for the "production" of these pathogens, of which it is a close relative.

What exactly is it? Ask the reader. Well, this is another concrete proof that a long time ago someone manipulated bacterial genes so that they would be a murderous weapon against the man and the animals that surround him. The lethality of these pathogens is against the strategy of survival of these bacteria, because pathogens care that their host lives as long as possible - in their own interest. For some time, the host's death was not in the interest of bacteria, which were perhaps even symbiotic bacteria with man and domestic animals. Due to the genetic manipulations described here, these symbiotic bacteria turned into pathogens and began to kill. The creators of this BMR have turned into dirt and dust for a long time, but their work is still collecting a bloody harvest even now - after one hundred and twenty centuries!

And here is the question that arises, namely: AIDS is a B weapon or a "gift" from the Kosmites or maybe Atlantis? It was set by Ivan Rybakov in the pages of the Russian magazine NLO:

Not so long ago in the English magazine "Steamshovel Press", devoted to paranormal phenomena, an article about the AIDS problem was published. Its author, Ken Thomas, examines several hypotheses about the origin of this strange disease. One of them talks about a biological experiment that got out of the control of scholars. Perhaps this was the result of the cold war arms race, when scholars tried to create the perfect biological weapon that would be able to destroy their enemy.

Secret presentation

The author refers to a secret report given by military expert Donald MacArthur in 1969 at a meeting of the United States Congress, during a meeting on the adoption of the budget for the next year. Here is a quote from this paper:

In the next 5 - 10 years, we will be able to grow microorganisms that will be able to affect the course of all known infectious diseases by significantly lowering the immunological abilities of the human body.

This is what AIDS is all about. In 1969, it was the peak period of the Cold War. Thomas thinks that at the time the group to which Mac Arthur belonged already conducted research into the production of immunodeficiency virus. Unfortunately, Thomas does not give a source that gave him information about Mac Arthur's secret talk. However, this source - let's call it NN - today the pensioner, who was a military specialist at the time, dealt with calculations of the speed and range of falling of a gas cloud contaminated by viruses. He does not know exactly what viruses but he is convinced that he is talking about viruses. NN is a believer, and when the AIDS evolved into a pandemic in recent years, he decided to tell the journalist what he knew.

Thomas decided to conduct a special journalist investigation to reveal those scholars who worked on this problem. He failed to collect direct evidence, but there were strong indications. Among others he talked to Mary Bagstock, whose husband with medical training, in his time worked as a lab technician at the United States Army Medical Research Institute of Infectious Diseases - US Medical Institute of Infectious Diseases Research - USAMRIID. Mary did not know exactly what he was doing there, but she knew that it was related to infectious diseases.

In 1972, Robert Bagstock fell ill unexpectedly and soon died. Doctors were unable to make a diagnosis because they detected several diseases, including pneumonia. Robert had a total of 27 years, and before that, he did not get seriously sick.

In 1983, Mary was suspected that her husband's death was related to his job and filed a lawsuit against the US Department of Defense. An exhumation was carried out, the result of which was unexpected because it turned out that the cause of death was HIV - Human Immunodeficiency Virus. Therefore, Robert died of Acquired Immune Deficiency Syndrome or AIDS. Mary failed to blame the US Army for the death of her husband. However, the very fact of death due to AIDS in 1972 says a lot and the facts of death as a result of HIV infection were not registered until the late 1970s! So, it looks like the HIV virus came out of the silence of the laboratories and spread to the world, hitting not only the military opponent but also other people - including its creators. Ken Thomas is convinced that similar research was also carried out in the USSR and other countries, and the dimensions of the thesis allow to believe that this was the case.

The AIDS and ETI hypothesis

The second hypothesis about the origin of the HIV virus is far more fantastic. It assumes that this biological bomb was brought to Earth by some aliens.

This view is shared by a whole range of ufologists, in this number an American specialist in researching the possibility of extraterrestrial Reason - Walter Maxlee. According to his views, they did it completely by accident. The HIV virus is not dangerous for the Kosmites themselves, because their immune system is immune to them. With Earthlings it is like with Eskimo, which has been transferred to Africa - dying of diseases that are harmless to Africans, and vice-versa - Negro transferred to Greenland will soon get sick and untreated - it will die.

The French ufologist - Gerard Bernier - is a supporter of another possibility, namely the deliberate HIV infection of humans. The goal is simple and clear - freeing the Earth from human domination and taking over the planet for yourself. HIV infection occurs through people taken on decks of UFOs. Bernier founded a dossier of people kidnapped by aliens, mainly Europeans. The Ufologist conducted a series of regressive hypnosis sessions with many of them, as a result of which they reminded themselves that the aliens aboard UFOs took blood from their veins and made some manipulations with it.

Mirelle Doufrane, a 60-year-old resident of Lyon who spent several days aboard NOL in July 2002, remembered that she had seen her blood in the butt placed in the center of some shiny circle, at which time a steady howl sounded. Then the same blood was brought back into the woman's blood system. The hijackers had the shape of an insect - in the ufological nomenclature they are called insectoids.

The second kidnapped woman, Claire Clavier, 24, told the ufologist that they were kidnapped by aliens very similar to people who were interested in how people merge in couples to breed offspring. According to Claire, they were very surprised that on Earth people are born due to the sexual intercourse of two sexes. There are also men and women alike, but children give birth to men after pairing with representatives of their sex.

Black perspective

Claire Clavier's statement from the French ufologist was considered crucial. It is known that HIV primarily infects homosexuals. Apparently, extraterrestrials have recognized that humans reproduce as they did in the result of same-sex acts. The most convincing evidence for the AIDS and ETI hypothesis is the fact that most of the UFO abductees have been infected with HIV during these CE4. Mirelle Doufrane died of AIDS in January and Claire Clavier in December 2004. Both women were respected in their environment, had no contact with drugs and did not maintain any sexual relations with other men. The diagnosis itself is astonishing. The French ufologist, on the other hand, has no doubt that the aliens infected both women with the HIV virus. Statistics speak for themselves - 12 of them have been kidnapped and 5 have died of AIDS and 1 person is HIV positive.

So what is AIDS? - Weapons B developed in secret military laboratories or maybe a "gift" from Przybyszów from space? Ken Thomas believes that both of these hypotheses are likely. Either way - the genie came out of the lamp - and if the medics do not find an effective method of treatment, the intelligent life on Earth will simply perish.

When reading the article by Ivan Rybakov, we came to the conclusion that it is worth presenting two other theories of the origin of AIDS. The first of them assumes that the HIV virus is a mutated descendant of some pathogen, which was created in natural conditions - Richard Preston indicates here a group of islands Sese (Uganda) on Lake Victoria, from where it later spread along the Kashoshka main (to Mombassa in Kenya to the east and to Pointe Noire in Zaire) to the west. Similarly, there was another deadly pathogen - the Ebola-Kenia virus, whose source is the Kitum Cave in the Mt. Elgon (4321 m above sea level). In the case of Ebola virus, it can be said that it meets all the conditions of the ideal weapon B - it is a 4th level virus (according to the USAMRIID nomenclature) and its lethality reaches 90%. Although the HIV virus has a lethality of 100%, its low contagion makes it only a level 2 virus. (=> R. Preston - Pollution Zone, Warsaw 1996, op cit. Pp. 21-80)

The second hypothesis assumes that both of these viruses have been mutated and used in the Atomic War of the Gods 12,000 years ago, and people have stumbled upon some contaminated containers or unexploded ordnance in the forests of Africa and South America, from where they re-emerged. This hypothesis is attractive in that it explains the facts without resorting to difficult or almost impossible to verify whom the aliens could modify the memory so that they would talk about various nonsense, as cited above by Rybakowa. One thing is for sure, strangers are interested in our reproduction, but this is due to their own problems in this regard.

However, many data indicate that the HIV virus can be a mutated virus of an animal disease that broke through the species barrier and attacked humans. An example from recent days is the epidemic of the so-called bird flu caused by the virus designated H5N1 in Asia and now also in Europe. Therefore, it cannot be clearly stated that HIV has fallen from the sky.

Another aspect of this issue. One of the hypotheses explaining all the turns and twists of human history is the hypothesis of Atomic Gods 'Wars - Astronauts - AWBA, which became an alternative to the hypothesis of Visitors' from Space and in general Daenikenism as such. In short, it assumes that there have never been any Visits in human history, and any anomalies in history can be explained by the fact that our civilization is only one of many human and pre-human civilizations existing on our planet for at least several million years. Personally, we share this view but not completely, because we allow the possibility of rare visits of the Przybyszów from the Cosmos at a frequency of once every 10,000 to a million years which we consider to be extremely optimistic. The huge distances in space and the multiplicity of stars in the galaxy (which includes our sun) in the galaxy makes the probability of discovering the Earth by the aliens is small but nevertheless, it exists and can happen at any time - even now, when we write these words .

Traces of AWBA in the micro world

We have already written elsewhere about the microscopic traces of AWBA and we will not repeat ourselves. It is enough to mention that the origin of such terrible delights of God, as plague or cholera indicates that in the not-too-distant past - estimated by biologists for about 20,000 years - unknown. Someone manipulated the genome of these bacteria and caused that from benign creatures they became infectiously malicious and ruthless exterminators of humans and animals on Earth. And people who have never been interested in the AWBA hypothesis and have not heard about it!

Not just a B weapon

But not only was the B weapon in use, because weapons A - nuclear and thermonuclear were used against people. There are numerous traces of it in the form of gravelified sands and stones in many points of the Earth. According to some researchers, they correspond with strange legends about the old gods and "cities of lights", which were located where today sands and gravels of large African, Asian, Australian and American deserts are pouring.

Popular science books and school textbooks for Defensive Education tell us that the operation of a weapon is so lethal that in the case of its massive use, life on our planet can be completely eliminated. Of course, AWBA did not completely eliminate life on Earth, but it could have caused a number of mutations and another episode of Great Deprecating - for example, the tertiary mega fauna. Thus, it did not have to be a lethal effect of the asteroid, a supernova explosion, a super-volcano or the greenhouse effect, and the polarity of the Earth. It could have been a total war of everyone with everyone and against all of them with the same effect.

Another trace of the use of WMD from the A weaponry arsenal are the tektites. They are strange glassy droplets of molten matter, whose fields are found all over the Earth. Their chemical composition departs significantly from the chemical composition of "ordinary" meteorites, and the low content of water in them makes their origin and nature very mysterious. The tectonic field closest to Poland is located in the Czech Republic and Slovakia - these are the so-called Moldavites. I refer the interested parties to the works of Andrzej Kotowiecki from the Polish Meteorite Society, whose articles about the tektites appeared in the pages of "Nieznane Świata" and "Meteoryt". Tektites was most likely created under conditions of high vacuum and temperatures in thousands of Kelvin, because they bear traces of the temperatures generated during fusion explosions.

However, these are traces found in inanimate nature. There are traces in the lively nature, which suggest that once there was a conflict on the Earth with the mass use of WMD, including weapons A, Aleksander Mora states that studies of scientists from many countries show that animals living in the deserts are much more resistant to radiation, than living in other conditions. The same applies to plants. This can be explained by the increasing resistance to irradiation due to the inhabited zones of radiant contamination. It could be a clear circumstantial evidence.

Radiotolerant bacteria

All these plants and animals are removing creatures from the micro world. It is commonly said that after World War II, only cockroaches and rats will remain on Earth. True, both species show increased resistance to irradiation but there are organisms super-resistant to radiation on Earth. They are bacteria from the species Deinococcus radiodurans - it is a bacterium that can withstand radiation doses that can be deadly to humans. It was discovered in the fifties of the twentieth century (exactly in 1956) in spoiled canned meat, which were reportedly subjected to gamma radiation for sterilization. Deinococcus radiodurans growing colonies were found in the meat with unique features of quickly rebuilding all genetic defects caused by radiation. These features are the source of the name of microorganisms, which can be explained as: foreign, radiation-resistant organisms. (=> "Mars on Earth" part 2 - www.marssociety.pl)

They were discovered by Dr. Arthur W. Anderson from the Oregon Agricultural Experiment Station in Corvallis (OR). He noted that in the preserved meat there were red colonies of bacteria that could withstand irradiation with gamma rays of several Megarads - 1 Mrad = $10 \wedge 6$ = 1,000,000 rads. By the way, also discovered a whole range of bacteria that are resistant to all extreme living conditions - they are so-called, Extremophiles. D. radiodurans owe their properties to the rapid removal of chromosome damage caused by radiation and restoring them to the original state.

I do not need to say how important it would be to have such a skill by the organisms of all living beings on Earth - or at least by humans. (=> T. Lottman - " Deinococcus radiodurans " - http://www.geocities.com/ResearchTriangle/Forum/1416/deino.html) .

These extremophiles occur in several species. They are resistant bacteria, among others hot (thermophiles) - D. thermophilus and D. igniterrae, as well as D. geothermalis; radioactivity (radiophiles) - D. radiodurans and D. radiophilus as well as D. radiopugnans. They may occur (and occur) in the interior of nuclear reactors and installations of primary circuits of water vapor of nuclear power plants. Atomic radiation is not too bad for them.

(=> Uniformed Services University of Medical Sciences [USUMS] - http://www.usuhs.mil/pat/deinococcus/index_20.htm).

What does all this mean? D. radiodurans has been named Conan among bacteria because it is also relatively large. Her resistance is unbelievable. It's enough to compare this bacteria with humans. To kill a man, you just need to irradiate him with a dose of 0.5 to 1 krad. D. radiodurans can withstand a dose of 1.5 Mrad, and some radiophiles up to 5 Mrad! And that means that these bacteria are able to survive even an atomic explosion and this is not far from the zero point! This raises further questions.

(=> S. Sleuth - "Microbe!" - http://www.microbe.org/microbes/Deinococcus.asp) Of course, the immune capabilities of these extraordinary organisms increase when the bacterium is almost dehydrated and is in the form of a spore. (=> "Amazing Facts" - http://www.scifun.ed.ac.uk/card/facts.html)

Visitors from space or children of war?

So far, it has not been possible to completely explain where these bacteria came from in the Earth's biosphere. It is possible that they were created together with the entire micro and mega fauna of the Earth, but the thing is, species are not created just like that. However, they are created under specific ecological conditions - in this case, in a high radioactivity environment.

There are no natural sources of radioactivity on our planet with such a concentration of radiation, even in the natural nuclear reactor in Oklo (Gabon). Perhaps these bacteria are some mutants that arose from nearly 2,500 atomic explosions in the deserts of Arizona, Nevada, Australia, the French Sahara, Gobi and Takla-Makan, Al-Shan, the steppes of Kazakhstan and Orenburg, the forestless boundaries of Yakutia, polar deserts New Earth and Pacific atolls. It may have arisen in experimental or industrial nuclear reactors. One thing is certain - they were created and evolved there in extreme conditions of high temperatures and huge concentrations of UV and gamma γ radiation, and this is the argument that they do not come only from this Earth.

Such extreme conditions prevail on Mars - where the concentration of UV radiation is deadly for living creatures from the Earth (apart from the fact of the rare and poisonous atmosphere on this planet) and Venus, where temperatures prevail in the order of 450°C, a poisonous atmosphere composed of carbon dioxide and clouds sulfuric acid, exerting pressure of 90 atm - these bacteria are able to survive and such a hell! These bacteria are resistant to most poisonous chemicals and concentrated acids! They resemble silicon - silicon bacteria from the novels of K. Borunia and A. Trepka - "Space Brothers" (Warsaw 1957, 1986), which conquered our planet due to our oversight. Could such bacteria be a pre-process for terraforming these planets? Krzysztof Boruń and Andrzej Trepka have just described such a process. Silihomids, this intelligent silicon, transformed the Earth into its own fashion, adapting its conditions to the requirements of its own existence: lack of water, temperature of about 1.000°C, boiling lava, sulfur clouds and other volcanic exhalations - in a medieval image of Hells.

Is something like this already on Earth? It may have been in some places after the AWBA and in fact, we are dealing with mutants after the thermonuclear war, or the remains after trying to terraform other planets. An attempt that may have failed - after all, Mars is frozen to the bone, and we see the effects of the greenhouse effect on Venus. [16]

Can Deinococcinae be visitors from outer space? And why not? They could fall with cosmic dust from comets, where they lived in extreme conditions of low temperatures, vacuum and powerful doses of UV, gamma γ and cosmic rays. Most scholars now adhere to the vision of comets - carriers of life in planetary systems and perhaps some viral diseases – e.g. flu, may be just such a "landing" from space. Research on these newcomers may help us to create effective

---

[16] This topic was developed by R. Leśniakiewicz in the work Fri "UFO and Time" (Tolkmicko 2011) in which he claims that this interest stems from the plan of restitution of the species Homo sapiens after the destruction of the biosphere by our planet.

protection against ionizing radiation and UV radiation, and I do not need to say what would be a step forward in the fight against radiation sickness and some types of cancer! Genetic research of these bacteria can help us to move forward in the development of nuclear energy, which will become more secure in the sense that it will be possible to protect living creatures against lethal effects of nuclear radiation in the case of such nuclear disasters as in Three Mile Island EJ or Chernobyl's EJ.[17]

We believe that the progress of genetic research on the micro flora will allow us to find more evidence that the manipulation of life took place much earlier than we think, and that now we pay tribute to the madness of the White and Black Magicians of Atlantis.

---

[17] And now also after the Daiichi-Fukushima 1 EJ disaster.

CHAPTER V

Shamballa and the Polish question

On Sunday, January 7, 2001, Robert Leśniakiewicz took part in the program "Nautilius Radio Zet" on Radio Zet, whose theme was the existence or non-existence of the mythical land of Shamballa-Agharta, also known as Realm D'Bus [read: Ü or Ui] or Shangri-La - as in James Hilton's novel "Lost Horizon". The program was also attended by Igor Witkowski - a well-known writer, author of a multi-volume work entitled "Hitler's secret weapons", "Super weapons of Islam" and books on the subject of UFOs, and a UFO researcher and the phenomenon of crop circles - Michał Zawadzki . All participants presented arguments pro and versus the existence of this strange underground world, which according to Buddhist messages and religions Bön, exists for 60,000-6,000 years! Here is what a Russian journalist Wadim Konstantynowicz Ilin from St. Petersburg writes about Shamballa-Agharta:

In the circles of the initiated people, the hypothesis that our Earth is empty inside is popular. The followers of this hypothesis have already lost (and are still losing) plenty of time and energy to seek to enter these underground spaces. Some think that such entrances should be near both Earth's poles, mainly beyond the northern Polar Circle and others hope to find them in the mountains of Central Asia, especially in Tibet. Currently, the search for "hidden doors" is conducted in a variety of ways, including using the latest scientific methods of collecting and processing information - from photography to satellite remote sensing and seismic surveys.

In earlier times, the basic sources of information to search, along with the results of polar research, were old manuscripts and oral translations. In one of the sacred books of Hinduism - "Bhagavat Purana" - one speaks of the legendary Maharaja Saghara - the mythical Emperor of the Indians, whose countless sons were known for many undertakings. Such a venture was also the search for a sacrificial mare devoted to one of the most powerful Hindu deities - the thunderous Indra. At the command of their father, they went to look for this horse, and searched the whole empire and the whole Earth. Not finding her, they went north and went inside the Earth, where they found this mare and a sage named Kapila. In other "Puranas", we learn a little more detail about this underground expedition. It is said in them that the sons of Maharaja got to the shores of the North Ocean, passed them and thus got into the Earth.

One of the most enchanting and at the same time the most mysterious legends of the East - the legend about the state of Agharta (Agharti, Agarti) - is in harmony with this ancient account. It stretches out in the depths of the Earth, and the entrance (or even a few entries) to it is somewhere in the territory of Tibet. As it is written in ancient Indian texts and stories of Tibetan lamas passed down from generation to generation, Agharta is inhabited by the descendants of the former Earth civilization - much older and much more technically advanced than our contemporary, and especially about its spiritual development. The people of Agharta live in peace and observe the fate of our world, which the threat of catastrophe is inexorably approaching. The critical moment will come in 2425, when the Third World War may break out but if you believe the messages,

Earthlings have a chance to avoid danger with the help of Aghartans and their ruler who wields the most terrible weapon in human history - Vril energy - the forces of Good overcome Evil and as a result on Earth, there will be the prosperity that exists in Agharta.

The legend about the wonderful land stimulated the human imagination at the turn of the 19th and 20th centuries. Who would not want to be in Paradise in this lifetime!? Many sought to enter Agharta, in this number: a French diplomat and writer living in India in the nineteenth century Louis Jacquliere, a Russian traveler and researcher of Central Asia Mikołaj Michajłowicz Przewalski (1839-1888), painter, writer and traveler Nicholas Konstantynowicz Roerich (1874-1947) and his son Jurij - orientalist and philosopher who lived in India for 20 years, as well as Polish explorer and writer Ferdynand Antoni Ossendowski , and on the order of Adolf Hitler himself, SS-Standartenführer Schauffer - a member of the leadership of the Nazi Institute for the Study of the History of the Spirit.

Extensive and accurate messages about Shambhala (Śambhalli, Shambhala) - Agharta gathered during the years of life in India Mikołaj and Yuri Roerich. The Buddhist beliefs speak of the "Eight Immortals" - eight Masters who inhabit the interior of the mountain on the border of China and Tibet. This place in some legends is called Agharta, while in other Hsi Uong Mu, according to many people is also located underground, not far from the capital of Tibet - Lhassa.

About eight Immortals and their residences in the mountains, Mikołaj Roerich heard during his expedition to Asia in the first decade of the twentieth century. From the guide who was the local inhabitant of this land, he learned that inside the mountain massif of Kunlun there is a huge cave with high vaults, where there are still treasures to this day. This guide also mentioned some mysterious "gray people". In contrast, the lama-prior of one of the Buddhist monasteries told Roerich the old Thai application about how the

Immortals were made of clay and air by Mu Kung - the ruler of the East air and Uong (Wong) Mu - the ruler of the Western air. However, according to later reports, Immortals appeared on Earth from the planet located in the star system of Sirius, and it was in the mountains of Tibet that they set up their outpost in order to carry out experiments of genetic hybridization.

In one of the books on Tibetan expeditions, Roerich announces that he saw a flying disk in the sky - according to modern terminology - NOL. The guide told him that such disks are arriving from Agharta. According to the beliefs of Tibetan Buddhists with centuries-old traditions, Agharta has a second, even more mysterious city - Shamballa - just like the Vatican is inside Rome.

During his trips to Tibet, Nicholas Konstantynowicz Roerich and his wife, Jelena Ivanov, talked with representatives of the Buddhist clergy on the above topics on many occasions. Some of these conversations are cited in Roerich's books, such as 'Altai - Himalaya' (1927), 'The Heart of Asia' (1929) and 'Shambhala' (1930).

I remember - he writes - as during our passage through the Karakorum mountains, my guide and assistant Hakaki asked me:

Do you know why there is such an unusual, mountainous region stretching ahead of us? Do you know that there are huge treasures in the underground caverns, and that there is an amazing nation in these caves that rejects all that is sinful on this earth?

I also remember that when we got closer to the city of Hotan, the impact of our horses' hoofs on the ground became deaf, as if literally below us were some caves or other empty underground spaces. Also other people who traveled in our caravan also paid our attention to this. When we saw the entrances to the caves, our caravan guides said:

"People used to live here, long ago, but now they all went to the Inner Earth. They found the entrance to the underground Empire.

Here is a fragment of the dialogue that ran between Roerich and one of the Tibetan lamas in 1928:

Roerich: Lama, tell me about Shambhala.

Lama: After all, you people of the West do not know anything about Shambhala, and you do not want to know anything. Surely you ask me simply out of curiosity and you will not understand this holy word.

After long persuasions and conversations, and the questioning and swearing of Roerich, the lama agreed to continue this conversation:

Lama: Great Shambhala is located far across the ocean. This is the great empire of the gods. It has nothing to do with Earth. Why do people show interest in it? Only far to the north, and only in some places, you could see the shining rays of Shambhala. The secrets of Shambhala are well hidden from outsiders.

Roerich: Lama, we know of the greatness of Shambhala and we know that this indescribably beautiful empire exists. We also know that some lamas of higher grades were in Shambhala. I know the story of the miraculous journey of a lama from Buryatia, how he was led through a narrow, secret passage. Therefore, do not tell me only about the heavenly Shambhala, but tell me about the one that exists on Earth - I know that there is also an earthly Shambhala. Tell me lama, how did this earthly Shambhala not be discovered by any of the earthly travelers? As you look at the maps, there are practically no white spots on them. Apparently, all mountain ranges have been plotted on maps, all valleys and rivers have been examined.

Lama: Well, as long as these people you call travelers, they have not found much on Earth. Oh, let someone try to get to Shambhala without being invited! You must have heard that streams and rivers flow from the high plateau, whose water is saturated with poison, lethal to humans. Perhaps you have seen people who died after inhaling the poisonous fumes when attempting to cross these rivers and streams. Many mortals tried to reach Shambhala without invitation. Some of them disappear without a trace forever. Only a few manage to get to the Holy City, and only in this case, when their karma allows them.

He devoted a lot of time and energy to finding Agharta during his retreat in Central Asia in the early twentieth century, the Polish traveler and writer Ferdynand Antoni Ossendowski (1878-1945). In his book, "Through the Land of People, Animals and Gods" (Poznań 1923), he describes his multiple meetings and conversations with lamas - Buddhist monks - of various ranks, as well as the simple inhabitants of these places, as well as the adventures associated with this search.

Ferdynand Ossendowski tried to unravel the mystery of Agharta over 80 years ago. The information provided by Mikołaj Konstantynowicz Roerich is younger by 10 years. Well, what do modern scientists of this legendary underground world of all welfare know and think about Agharta?

The number of today's problem researchers is Ian Lamprecht, who in his book entitled, "Hollow Planets" (1998) describes that in the American city of San José, CA, he lectured a lama, a doctor of Tibetan medicine, a scholar and teacher of one of the directions of Buddhism - Vajrayana . His name and title are as follows: His Holiness Orgen Kusum Lingpa. There are some clues to believe that His Holiness belongs to a very narrow circle of people who have much to say about the esoterics of the East, including the topic discussed here. Here is what he writes about this in his book Lamprecht:

During one of the lectures given in San José, that lama said that you can get to Agharta flying from India to the north in seven days. I am sure that the lama was about a journey with a medium speed of a bird's flight, and if so - a seven-day flight at such a speed will take us to the very center of the Arctic. Seventy with a wrap up years ago, Nicholas Roerich talking to another lama heard from him that Shambhala lies far to the north. So, perhaps this lama was thinking of the Northern Icy Ocean?

Dr. Ludmiła Szaposznikowa, who was in India, sees this differently, where she came across the confessors of the pre-Buddhist religion of Bön. In her reportage published, among others in the pages of "Nieznany Świat" No. 11/1994 she writes about the legendary land. According to religious scholars of the Bön religion, in the manuscripts of believers of this religion, it is clearly said that the sky, the earth and the feast of Shambhala were created. European scholars argue that it is in the mountains of Kajas. Its interlocutor - rimpoche Senge Tensing Iongdan claimed that:

Shambhala is located in the region of the western Himalayas, somewhere between the western part of the Indian Himalayas, Lhassa and Transhimalaya. A long time ago in this territory was the Shan-Shung state, which was divided into three parts: Szan-Szung Bu, Szang-Szung Par and Szan-Szung Go, the latter occupying the territory north-east of Kajas to the monastery of Bön called Czun- Po. Szan-Szung Par stretched west of Kajas and bordered with Afghanistan, while Szan-Szung Bu was a saint and blessed Oma-mo-lun-rin, or Shambhala , she writes.

And next:

After entering Szan-Szung Bu through the Tang-la pass, he descends to a wonderful land, the first harbinger of which is the wonderful tree Pe-dum Dum-po with red flowers. Next are the gardens and the lake.

Step by step, going north-east, we were approaching the place where the Central Tower stood. We stood by the columns rising into the infinite sky. They were made of monolithic blocks of mountain crystal, and their transparent planes covered some writing.

"Is this a letter from the land of Szan-Szun?" I asked Rimpoche.

"Yes," she confirmed, "it is szan-szun, the old language of the gods, a long-forgotten letter. Only protected on these columns.

We hurried to the tower, which was raised above us, built of large blocks of rock. We stood in front of it. The tower looked at us through the wide windows in the stone walls.

"This tower is something most important on our planet," said the rimpochev, "it has many hundreds of millennia, and who knows, or not a million years. In this tower there is a hidden Mystery. Only the wise men of Bönu - the Siddhas - could enter her and even few. I do not have the courage to go there with you.

"It is a pity," I thought, " to be so close to the greatest mystery of the planet and leave with nothing."

"It is a pity, great pity," he repeated the rimpochets for me, "but time will come. We must return."

So much the relation of Dr. Szaposznikowa. As you've probably noticed, the reader is very similar to what James Hilton described in the novel "The Lost Horizon" and what was shown in the movie with the same title. We find a similar description of the mysterious country in the last book from James Redfield's trilogy.

For Robert Leśniakiewicz, this is a family matter, as his grandfather Franciszek Baranowicz spent a dozen or so years in Siberia and northern China, where in 1905 he fought in Manchuria with the Japanese. Mr. Baranowicz also came across this legend. The Polish author, who in the twenty years between the wars propagated it in Poland - a writer and traveler - Antoni Ferdynand Ossendowski, wrote about it in the book "Through the Land of People, Animals and Gods". He himself, since 1980, has been collecting all information about this underground land, which - assuming Ossendowski's opinion - exerts a significant influence on what is happening on Earth, and some indications suggest that the phenomenon commonly known as a UFO has its source there - in Agharta. Today we can say that it was not true - in the case of Babia Góra - about UFOs, but about the terrestrial vehicles manufactured in the Third Reich, thanks to which Hitler managed to escape from burning Berlin to the Chancellery of the 4th Reich on a rocky the coast of the Peary Land in Greenland. He wrote about it in the pages of "Nieznany Świat" and on his website.

At the time of his presentation he spoke about the Polish part of the Agharta Legend, where the alleged entry was supposed to be. If you believe a well-known Polish scholar employed at Aoraki Polytechnic and Dunedin University in New Zealand, Prof. Dr inż. Jan Pająk – the entrance is to be located no more than on the south-west slope of Babia Góra, somewhere at the height of 2/3 of the relative height of the Queen of the Beskids - that is, at an altitude of 1550 - 1650 m above sea level, on the Slovak side of the border. According to Prof. Pająk, a legend about corridors in Babia Góra, which supposedly reported not only to Shambala-Agharta, but also to America and China, he heard from an inhabitant of one of the Beskid villages who entrusted him with a secret. The entrance was under a group of rocks on the Babia slope, and the tunnel itself was several meters wide and its walls were covered with a kind of glaze, glazed with very high temperature rocks. The corridor led to a spacious barrel-shaped chamber where, according to the informer, Prof. Pająk - wait even for the biggest cataclysm like thermonuclear war or natural disaster. Was it a real relationship, a fairy tale for tourists, and perhaps an echo of legends about the Mooncave, whose creators are Agartians or Atlantians?

CHAPTER VI

Australian Atlantis

This article by Ivan Rybakov appeared in the pages of the Russian journal NLO, No. 50/2005 dated December 12, 2005. It is controversial, but only at first glance, because after a long reflection you can come to the conclusion that something is in it.

World missing on "Turkey"

The remains of the "lost continent" were discovered on the bottom of the Indian Ocean at a distance of about 2,500 miles (or about 4,000 km) south-west of the Australian port of Perth. It turned out that the Heard Island and the MacDonald Archipelago are in fact the tops of the colossal land mass, which is now under water. Long ago, during the Cretaceous period, this huge island or small continent was covered with lush thickets of ferns, huge reptiles and dinosaurs followed him.

This "Lost World", whose surface occupies ⅓ of the area of present-day Australia, has risen from the bottom of the ocean between 90 and 115 million years ago. A series of above-ground and underground explosions began a terrible cataclysm. Over the next 20 million years, this piece of land drifted and gradually decreased, and after another terrible cataclysm, it finally sank into the bottom of the ocean.

Continental rock structures

Australian scholars have come to this conclusion on the subject of the old continent after a study in the summer of 2005 on the Kerguelen Island, an underwater mountain ridge that divides the Australian-Antarctic Basin from the African-Antarctic Basin. With the help of special devices, they were able to take at six points samples of rocks located at 1 km below the surface of the ocean floor. This took place on the Kerguelen Plateau, where the depth of the Indian Ocean is only 73 meters.

Oceanographers working on the research ship RV Resolution, concluded on the basis of the types of excavated rocks, that they are effusive type rocks, i.e. they were formed as a result of volcanic activity and long ago liquid magma flowed here. These rocks are exactly the same as the continental structures of Antarctica and India. In addition, in the samples of rocks excavated on the surface, traces of soil containing fossil plant remains in the form of stone and coal were found. These samples were tested at the University of Sydney's laboratory. It turned out that these are the remains of trees and woody ferns.

Further research of Australian Atlantis

This important discovery of Australian geologists and oceanographers has allowed extending the scientific insight into the past of the Cretaceous period. These were geologically restless times when volcanic eruptions carried huge tsunami waves from the bottom of the oceans. The earth was rising and falling. Today's lands have not yet been. Two huge supercontinents: Laurazia and Gondwana separated the Ocean of Tetida (Tetys). These continents were fragmented by tectonics and magma explosions. The southern part of North America broke away from Laurazia, the Decan Peninsula (present-day India), and then Antarctica with Australia and South America sailed away from Africa. Today's oceans have formed and a small continent has appeared, which until recently was not known to geologists. Head of the Australian expedition to Kergueleny, Prof. Mikel Coffin proposed to name this continent of Australian Atlantis. This year, an international expedition is planned to be attended by Australian, American and British scientists with the aim of further researching the 'lost continent'.

"How much is the truth about it?" we ask. It turns out that this missing world could really exist! And who knows if it was about him in his delirious visions wrote Howard Phillips Lovecraft (1890-1937) describing the underwater cities of R'Lyeth and the Antarctic city of black towers Kaddath, about which I wrote in the pages of "Unknown World". (=> HP Lovecraft - "Call of Cthulhu" and "In the mountains of madness") As the research of literary historians has shown, it often happened that the intuition and extremely colorful imagination of writers is able to create visions that coincide with facts to a large extent. In such cases, people are referred to the visionaries, such as Stanisław Lem's (1921- 2006).

The bathymetric map shows that, indeed, the islands of Kerguelen and the whole MacDonald Archipelago are located on the undersea elevation, which perfectly overlooks the plastic map of the bottom of the All-Ocean. Once, it could actually be quite a large land with its own unique environment. Who knows if there have been any survivors of dinosaurs after the great catastrophe caused by the impact of the asteroid and the possible gigantic explosions of the super-volcano under Decan and Indonesia? There are many indications that this mysterious land could have been such an oasis of life that survived even the post-active and / or post-exploration winter. I would not be too surprised if the tertiary sediments of the leinazaurs and common driosaurs were found in the southern hemisphere in Upper Kreis. The Kerguelian Plateau probably stretched above the ocean level 12,000 years ago, when a large amount of oceanic water was trapped in glaciers in the Northern Hemisphere of our planet. If there was a civilization of Atlantis, there could have been a colony there, which after melting the glaciers shared the fate of the Atlant Empire during "one day and one night of horrible," as Plato described it colorfully. Anyway, such "phantom islands" is more and accurate analysis of the relief of the bottom of the All-Ocean should provide us with more data about their location. And for now, I am waiting for the results of this latest expedition in 2006.

But what about Atlantis! After all, there is an even more mysterious land in which descendants of the creators of the Moon's Cave can live. Natalia Kowalewa writes about her, the article we quote here. Prayers about the mysterious land of spiritual guides and sages, existing somewhere in the East, meet in the folklore of many countries and nations of the world. What is behind these legends? Is this just an extraordinary myth vitality that is widespread in many cultures of the world? Or maybe these are only relations coming to us from the depths of Time that talks about something that existed centuries ago and now circulating among the nations of the world? And here is the material from the Russian magazine "Kalejdoskop NLO" No. 16/2007, referring to the report published a few years ago in the "Unknown World" by Dr. Ludmiła Szaposznikowa titled "Shamballa dawna i tajemkowa":

Shamballa, this name for many people is associated with the existence of some Paradise, Garden of Eden or yet another land of eternal happiness and Ultimate Wisdom. There were people who were looking for this Paradise on our planet thinking that it is a specific geographical region. And one of them is mentioned in this article by Konstantin Koltsova published in the Russian journal "Kaleidoscope NLO" No. 31/2007.

An unusual CV

Mikołaj Roerich. When we mention this name, most people associate it with beautiful pictures. Nikolai Konstantinovich was a very talented man who was able to manifest himself in various areas of life and art. A painter, scholar, philosopher, cultural activist, writer and traveler. First of all, Roerich himself considered his greatest achievements to be his travels to Central Asia. He himself divided his life into two stages: preparation of an expedition and preparation of its results.

Nikolai Roerich was born on October 9, 1874, in St. Petersburg. Roerichs' protoplasts (from Norway) appeared in Russia in the eighteenth century. Interest in the affairs of the East and Central Asia came to Roerich very early. There was a good atmosphere at home for this. The youth often spoke in historical, literary and scientific discussions about the East.

After 1900, Mikołaj Konstantynowicz together with his wife Helena Iwanowa de domo Szaposznikowa set about studying Hindu culture and philosophy. In 1913, he published an article entitled "Indian Way", in which he wrote about the importance of researching the culture of this country and organizing an expedition there. Well, it is obvious that as a true scholar, he could not rely only on book searches from behind the desk. Just before the start of the First World War, together with his friend, archaeologist Gołubiew, Roerich begins to prepare for the journey, which aims to explore the original sources of Eastern philosophy and ancient monuments of culture.

After the communist coup

Well, in 1917 a Bolshevik revolution took place, which changed the fate of his homeland. This Renaissance artist understood that Russia would not go to India due to the civil war. So he first goes to Finland and then goes to Great Britain. And it was in the capital of the British Empire that he heard about Shamballa for the first time from the Russian theosophists. Old legends said that somewhere out there, in the mountains, on the border between India and China, there is Shamballa - the seat of gods and a repository of ancient knowledge. To find this "Paradise on Earth" attempted many travelers over many, many centuries.

It should be noted that Roerich immediately believed in the reality of Shamballa's existence, for which he found a suitable place on the map - the north-western part of the Tibetan Plateau. In addition, he supposed that it is there that man can acquire knowledge allowing for the harmonization of internal and external energies, thanks to which he will achieve a higher degree of cosmic consciousness. Besides, this Russian was going to create a huge Mongolian Siberian state - New Country within the Asian continent.

(Similar dreams according to Antoni Ferdynand Ossendowski [1876-1945] were alive in "Nowy Gengiz-chan" general-lieutenant baron Roman Nicolaus Fiodorowicz Ungern von Sternberg [1886-1921] - commander of the counter-revolutionary Asian Horse Division and other counter-revolutionaries - crowd note)

His system should be based on the Buddhist worldview and should be the seat of the living embodiment of the Buddha. In other words, the materialization of the Northern Shamballa in the world would be made.

First, you had to find evidence for this in the search. Unexpected help came from America. The Yankee patrons gave the Russian painter so much money that he could not only organize the expedition but set up his museum in New York.

In December 1923, the Roerich family began a large expedition to Shamballa. The first stop on their way became the tiny mountain duchy of Sikkim, where a long-awaited meeting with the Tibetan lamas took place. The monks handed him a message from the rulers of Shamballa, which was to be handed over to Włodzimierz Ilicz Ulianow Lenin (1870-1924) as soon as possible. The painter stopped his journey and went to Europe, where he learned about the death of the "proletarian leader".

Expedition

Move forward with the expedition managed in August 1925. Under his leadership, there were 50 people and 80 pack horses. And so, this caravan managed to pass the first mysterious circle in Central Asia on the caravan trail leading through Punjab, Kashmir, Little Tibet, Hotan, Jarkend, Kashgar and Turfan. The travelers had to cross the high mountain passes and rushing mountain rivers and through the areas of various tribes - not always peacefully oriented towards newcomers. Already the first snow storm - a typical Siberian burania almost put a cross on their plans - Roerich and their companions only thanks to this escaped with life, that Tibetan horses have found their way in a raging blizzard.

Then a mysterious Mongolian lama joined the expedition - speaking directly in Russian - a GPU agent (Soviet secret political police, later KGB) Yakov Bliumkin planning to get to Tibet and start a revolution there with Roerich.

(Ossendowski, mentioned here, describes in his novels "Through the land of people, animals and gods" and "Lenin" this kind of red agent activity in Manchuria and Mongolia, whose aim was to destroy the base for counter-revolutionary activity in Siberia and invoke communist upheavals October).

This fact came to the British intelligence, which was given the task of stopping the "red spies". In October 1925, at the behest of a British resident in Tibet - Bailey, Roerich was arrested in the Chinese town of Sińczin and was forbidden to move on. After a six-month wait, the caravan moved north and crossed the Soviet border.

A message from India

In July 1926, Roerich visited Moscow and handed the Messenger to foreign affairs Chicherin a message from the mahatmas (living Buddhist saints).

This hardened atheist, with a smirk of irony, read the holy text of the message, and strangely, provided comprehensive assistance and support to the traveler whose path now led to Altai.

The next expedition went to Mongolia. On the streets of Urgi (today Ulaanbaatar) the locals sang songs about Shamballa. Listening to them, the traveler understood that the country he was looking for was somewhere close. So they went to the deserts of Takla-Makan and Gobi. Mikołaj Konstantynowicz collected many such strange facts during this trip, in which his guides were local residents and lamas.

(There is also a second polonicum in this matter and we find them in one of the books by Alfred Szklarski [1912-1992] - "Tom's mysterious trip" in which the author quoted the Mongolian legend of the land of Ui or Üi, under which the underground sea was supposed to be. The name Ui is the ancient name of today's Tibet - crowd note)

In one of the areas of the Takla-Makan Desert, north of the rocky peaks of Karakorum, the natives told him that "holy people live behind that mountain who save humanity with their wisdom; many people tried to see them, but they failed - as soon as they went higher up, they could not find a way. "

A Hindu guide told Roerich about the existence of huge caves in the Karakoram mountains, in which treasures were collected from the very beginning of history. He also claimed that white people of high height were seen there, who disappeared deep within these rock galleries. Roerich himself often found such oases in Tibet in such harsh climatic conditions, where their existence was simply impossible.

After the great expedition

Travelers walked long forgotten paths, next to ruins forgotten by gods and people of cities, still remembering the rule of Genghiz Khan . Nomads attacked them many times. But Tibet failed to overcome them: the Dalai Lama's government, under pressure from the English, did not let strangers into its land.

For five months, the expedition waited for approaches to Lhassa - the capital of Tibet - suffering from cold and unstable. People came to overwinter in the bare steppe, in icy winds that cut the blood in the veins with frost. Almost all horses died and half of the guides died. But Roerich did not stop his scientific work, even under such conditions. He puts dozens of mountain peaks on the map, collected whole assortments of animals, minerals and archaeological artifacts.

The Roerichs were sending letters to the Dalai Lama and the Governor all the time but he was receiving the same refusals in response. Finally, Nicholas Roerich gave up the idea of coming to Lhassy, where according to legend, the entrance to Szamballa was to be located. From that moment, Shamballa had nothing to do with Tibet for him - this "museum relic of the past".

On March 4, 1928, travelers returned back. This is where the journey of our great countryman ended. Roerich, as the first of Europeans, made the transition of Tibet from north to south through Transhimalaya and entering India.

After returning from Tibet, Roerich decided to stay abroad knowing that in the USSR, the WCzK (Soviet political police, later GPU, OGPU, NKVD, NKGB and KGB) eliminated all of his colleagues and friends in turn. In 1929, Mikołaj Konstantynowicz established the Himalayan Research Institute according to an official statement - "to develop the results of the expedition and subsequent research". The institute was located in the Nagar estate in the Kulu valley, the western Himalayas in India. Over time, the Institute became his residence. And that's where he died on December 13, 1947. until his last day, Mikołaj Konstantynowicz did not lose hope of finding the legendary Szamball. And he dreamed about returning to his homeland.

Prayers about the mysterious land of spiritual guides and sages, existing somewhere in the East, meet in the folklore of many countries and nations of the world. What is behind these legends? Is this just an extraordinary myth vitality that is widespread in many cultures of the world? Or maybe these are only relations coming to us from the depths of Time that talks about something that existed centuries ago and now circulating among the nations of the world? And here is the material by Natalia Kowalewej from the Russian magazine "Kalejdoskop NLO" No. 16/2007, referring to a few years ago published in the pages of the Unknown World by Dr Ludmiła Szaposznikowa "Old and enigmatic Shambhala":

Mysterious land

According to esoteric applications, Shamballa (also: Śambhallah, Shampullah or literary Shangri-la, there is also the Aghart name in Polish literature popularized by Antoni Ferdynand Ossendowski in the bestselling novel "Through the Land of People, Animals and Gods", Poznań 1930 - It is a land located somewhere on the border of Nepal, Tibet and India, where the adepts of the higher yoga are worshiped by holy teachers from previous eras. They had power over matter and other hidden forces of Nature already when our ancestors inhabited the caves. The inhabitants of this mysterious land were ahead of our civilization in terms of spiritual and scientific and technical development. Some scholars of the Szamballi phenomenon assume that the Himalayan Adepts Brotherhood consists of descendants of Atlantis and Lemurs who survived total cataclysms. But in the teachings of theosophy and Agni Yoga there are other testimonies. So, in Helena Pietrowna Bławatska's The Secret Doctrine, 1888, it is said that Shamballa existed in the time of Atlantis, and the members of the enigmatic Brotherhood of the Enlightened were not only witnesses of the loss of a huge continent, but also actively participated in saving every better representatives of the Atlantean civilization.

Szamballijczycy in the history of Humanity

As the scholars of theosophy and Agni Yoga say, the legendary land of the East was founded on our planet to accompany the evolution of Humanity and, above all, its spiritual development. Throughout the history of the Earth, members of the mysterious Brotherhood incarnated (they were simply born as ordinary people) in various countries around the world to contribute to the spiritual and intellectual development of the whole world. And so, representatives of Szamballa contributed to the development of all aspects of the culture of our civilization. It was just from Szamball that there were famous figures from our past, spiritual and political leaders of nations, genius creators of art and science, pioneers and explorers. To them, the esoteric tradition includes Paracelsus and Dante , Giordano Bruno and Pythagoras , Plato-Aristocles of the Athenian and Apolonius of Tiana, Tomasso di Campabella and Leonardo da Vinci , Akbar the Great and Ramses II the Great , Saint Germaine and Joan of Arc and many, many other mysterious geniuses who have left their mark in human history.

Previous epochs - which is obvious - brought various false, folklore-mythological and religious additional legends to Szamball's application. However, in the 19th and 20th centuries Szamballi phenomenon is exposed on a completely new level. Real information about Szamball traveled to the West thanks to the Russian members of the Adepts Brotherhood - the already mentioned Blavatsky, and the Roerich family behind them. And the works of Blavatsky and the teachings of Agni Yoga (that is, Living Ethics) have become the most complete and most objective sources of knowledge about the life and activities of the representatives of the White Brotherhood, as it is now called Szamballę w Zachodzie. Bławatska and Roerichi openly told the whole world that Shamballa is not a fairytale or someone's invention, but something real, while spiritual Szamballi teachers are real people of flesh and blood who, however, have extraordinary, paranormal possibilities, even according to our standards.

Upcoming changes

Why then did this mysterious, spiritual land of the East at once give its existence to the whole world and, moreover, gave the world treasured scientific and philosophical treasures available only for the Enlightened? Judging from all this, the spiritual Teachers of the East discovered the secrets of their existence to the world in order to teach practical lessons to people during the difficult times of the Transformation of Cosmic Epochs. Former prophets from the East and West unison spoke about the Earth's era of great cosmic changes that will be preceded by enormous natural and social cataclysms. Theosophy and Agni Yoga teachings contain even more important scientific information about the nature and causes of the events facing the planet. This knowledge will help people to enter into a new cosmic era.

The place of residence of this mysterious nation of the East is considered Hidden Valley in the Himalayas, which according to the applications is located in the area of the holy highlands Kanchendangi.

(=> James Redfield - "Heavenly Prophecy", "The 10th Initiation" and "The Secret of Shambhala", Warsaw 2000-2004 - editor's note)

On the mountain peaks surrounding the valley, there are snows and icy winds, while in the Valley located in the heart of the mountains hot springs beat - geysers, and the climate there is subtropical.

(It was very well shown by Charles Jarrott in the film Lost Horizon of 1973, based on the novel by James Hilton under the same title published in 1936, with music and songs by Burt Bacharach, and an older film directed by Frank Capra in 1937, also about Shangri-la. - note crowd.)

To this mountain valley, according to legends, perfectly masked underground passages and tunnels run from all sides of the Himalayas. Other times these underground roads begin in the mountain caves, sometimes in the underground catacombs of Lama's monasteries located in practically inaccessible areas of mountainous Tibet.

(Bod-rang-skyong-ljongs in Tibetan and Xizàng Zìzhìqū in Chinese, currently Chinese province since 1949).

Roads to the holy shamball ashrams can also lead through dangerous mountainous mountains and exposed cliffs, through shaky bamboo bridges over precipitous canyons and precipices.

In the foreground of Szamball

In the descriptions of travelers and traders' accounts one can read that when approaching the enchanted borders of the sacred land, people and animals felt unusual impressions, like hitting invisible rays. In these areas, caravan animals and people stopped at once and no force could force them to cross the sacred limits.

In the books of Nikolai Konstantinovich Roerich, it is written that in the Himalayas there is a huge number of caves, in which, according to statements of local worshipers, underground corridors start, which extend over great distances and reach Kanchendanga.

(Not to be confused with the himalayan Kangchenjung peak [K'ancz'endzönga, Kancz, Kanczendzanga] - 8586 m above sea level, located on the border between Nepal and India (Sikkim) in the eastern Himalayas, these are two different concepts and geographical names - crowd note)

These underground corridors lead to a beautiful valley hidden in the heart of the mountains. Similar descriptions are given in the works by another Asian researcher Prof. Ferdynand Ossendowski. During his journey to Central Asia, the Mongolian lama told him not only about the underground network of corridors and tunnels adjacent to the boundaries of the Holy Valley, but also about the existence of ultrafast means of transport in these tunnels - strange luminous apparatus that travel through these underground arteries. Dr Ossendowski, citing the Mongol lamas, claims that in some corridors and underground chambers there is a strange light in the light of which vegetables and fruits grow.

Towards the borders of Szamball!

During his Himalayan expeditions, NK Roerich often questioned Buddhist Tibetan monks for Szamball, recalling the legends about the underground corridors leading into the mountain valley, where the puzzling Brotherhood of Adepts resides. The lamas replied that while the Eastern Sages did not wish that they would be troubled by crowds of curious people from all over the world, this uninvited guest would never reach the temples of the magical land.

(On this subject, writes Jeremi Parnow in the novel "Zbudź się w Famaguście " in "Fantastyka" No. 11-12 / 1983, in which the hero arrives at Szamballi. Another opinion is Mircea Eliade [1907-1986], who in the short story "The Mystery of Doctor Hönigberger", Kraków 1983 claims that Shamballa is an extra-dimensional and oversized being and can only reach it spiritually, not the body itself.

Without a guide, you can easily die in a tangled underground labyrinth, where natural poison gases emerge from the gaps, which will surely kill the curious. In addition, there are other natural dangers lurking for a wanderer heading towards Shamball's borders. These are screens emanating energy unknown to Western science.

So Shamballa is not a myth, but something specific, but it is not available to our science? It seems that on Earth with us there is a second, highly developed civilization, which for its purpose of existence recognized the help in the evolution of the entire planet and humanity? The answer to this question is - I think very far ahead of us.

* * *

Comments of specialists

The above text would be incomplete without two comments from people who have been researching information about Szamball for many years. This problem is related not only to the problem of existence of Agharta and UFOs, but also to the atomic wars of gods-astronauts from thousands of years ago. The first of them is a resident in Malaysia, a French trader Patrick Moncelet from Kota Kinabalu, who writes about it like this:

Personally, I am convinced that Shangri-la, Shamballa and Agharta are just legends: if they really existed, Google Earth Project would have long ago detected them on the basis of satellite surveys. Many would say, "Oh! You're wrong!" After all, these are underground lands, but you can detect them (here I recommend a movie called "The League of the Extraordinary Gentlemen") based on their emission of pollutants or contacts with our world. I will give a simple example of the Badhuys tribe from western Java (Indonesia): they were discovered only because several members of the tribe decided to go outside Badhua Luar out of sheer curiosity or for economic reasons - some of them soon became wealthy people.

This tribe is one of the greatest mysteries of this world. I think you heard about Madame Bławatska ? Her nineteenth century theosophical movement is well known to me. I am still reading the book "Need to Know" by Timothy Good , which I recommend because of the new information from the American military about their UFO experiences. Mr. Good is doing a great job of collecting, developing and archiving information and interviewing many people, including former military and usually finds good sources. Like me, he thinks that most UFOs come from ultra-secret submarine bases that cannot be detected because they control their environmental pollutants in which they are found and produce their own energy, remaining hidden for several reasons.

Of course, I spent some time studying the matter of the atomic wars of gods-astronauts you raised, which were used, among others, Vimana vehicles. In Singapore's Little India, I spoke with a man who owns a small Indian bookstore with books in English and Hindi. He told me that yes, Vimana is an authentic Sanskrit word, but he has a whole range of meanings, but he did not know exactly what. Incidentally, I bought a Sanskrit dictionary from him intrigued by this language, a long time ago I had a friend who was fluent in it. Now there are entire websites devoted only to the problems of Viman and other vehicles. In Madras, there is a group of people studying this subject (at the local university), but they have many problems with translations from the old Sanskrit, because these texts are not as precise as it seems. One thing that is certain is that in the time of the Empire of Rama, there was astonishing metallurgical art.

Naturally, in the early days of Vimana, they were what UFOs are today - so who built them? One of the Hindu kings, I will not remember his name anymore, ordered to collect all Vimana , all technical documentation of their production and all sources about them, and then destroy them. Some researchers, however, think that several of them have been hidden in the mountains of Afghanistan, in caves and caverns. None of them has been found to this day.

You must have heard of the ancient city of Mohenjo Daro in India; its ruins are evidence of the use of nuclear weapons there, as well as in several other places in the Thar Desert, as well as desert glass (tektytach) in the Libyan Desert? Mr. Zacharia Sitchin writes about it in his works.

My Indian from Singapore is not very competent in this matter and he sent me to Madras, which is the cultural center of India, and where there are scholars fluent in this language and gave me their addresses. However, the translation of these old texts is an extremely difficult and thankless task: there are many possibilities for interpretation, because the language itself is very imprecise.

The descriptions of Viman and other flying machines of Antiquities are not technical or literary descriptions. Their drawings are not drawings from the original, but from descriptions and were made by Hindu and English translators who had no idea about the technique and the slightest idea of how these machines really looked. At present, in India, there are a few old people who know something of value on this subject (some yogis or swami), because these people have taught and taught their whole life and besides, they had nothing else to do.

When I was still living in Taiwan, at the end of the 1970s, I was presented to one of my sons - a teacher who was doing there his missionary work, creating an ashram there. He was very friendly, so after his stories about Swami Vivekananda I asked him: Swami, what do you think of "flying saucers" or UFOs?

He replied that "they exist," but unfortunately, he knew nothing more about them, but it encouraged people to study ufology. Even Maharishi Mahesh Yoga did not know more about UFOs than us, and even far less.

Speaking of yoga, I have a large reserve of yogis and their way of seeing the world and how they describe Reality. It is completely different from ours, and none of them - Sri Aurobindo, Vivekananda, Patanjali, etc. offered an explanation based on the true nature of the Time and the fractal nature of the Universe: they suggested the existence of prana and ether (Atma and Akash), but technically speaking no it is acceptable from our point of view who wants to know the ultimate truth of cosmology.

And this is Piotr Listkiewicz's answer :

Patrick is wrong in many places, because what yogini and swamis know is very little, as you see. Anyway, one can say for their excuse that they are not issues that they deal with every day, because their goal is completely different. Real swami does not travel to collect money (begging) to build ashram. The same ashram is created around the true Master. It is not a building or a housing estate, but people, students. Students gathered around Jesus and talked with him in the open air or at someone at home. Churches were created later.

Vimans were recreated not only on the basis of ancient descriptions. The description served as a prop to focus. He drew them a technician under the guidance of a man who guided him under the influence of internal insight. He saw these vehicles as in an x-ray picture.

Google's argument is naive. Of course, there are pages devoted to Shamball, and the map is missing, because nobody has yet located it. Do not overestimate the satellites - they do not see everything, and besides, some maps do not show details. Fragments of the terrain that is not to be seen by the general are blurred or do not have more detailed close-ups. I discovered it a long time ago, when I was looking for certain places, eg. Man from Maree.

Flying saucers, cigars, etc. are there for sure and this cannot be denied. Too many people saw them. They were often crowds, pilots, astronauts and scientists. As I wrote to you once, a Master said that they are partly natural phenomena of Nature, but some are artificial and belong to Earthlings. He described them as "aircraft of powers" without naming. However, because in the times when he was saying this (1960s) there were only two superpowers leading the Cold War, it was to be expected that over time these vehicles would be declassified and officially travel across the sky.

Meanwhile, over 40 years have elapsed and the secret has not been resolved. Think, if these vehicles belong to the USA and Russia, then the interviews of both superpowers perfectly know about them along with all technical details. So far, many new constructions have been created and none of the powers mentioned has any objection to boasting them. What extraordinary is it in the flying saucer that it is wrapped in such a heavy veil of mystery?

It is clear from this that there is still another power (the power?) On Earth, about which we know nothing officially. The US and Russian conspiracy are growing with each passing moment.

Disinformation is also growing, but some premises suggest that the ruling circles, and especially the military staffs of most countries, are afraid of something. What? Attention of researchers of these phenomena should focus on the moon. You know my opinion on this subject. The most suspicious is the sudden cessation of the Apollo program and the Moon's program of studying the Moon. At the same time. The Russians sent Łunochoda to the invisible side of the Moon and the data they captured never saw the light of day. Apparently, many of them were destroyed before they landed, and others later. What is on the moon, what are they all afraid of? If he was actually dead and uninhabited, there would be no reason to stop continuing both programs. Since then, the technique has multiplied its range and what was once only in the hands of the army of the most powerful countries, we have on our desks at home and in pockets. The moon was an extremely tasty morsel for those who would acquire it and develop it, because it gave power over the entire Earth and made it much easier to start exploring our solar system and further outer space. And suddenly, none of these two powers give up this goal?

However, if you do not look at it, the Moon is a dead globe - at least on the surface. What's inside, in the cave system? Caves must be, because the Moon has a volcanic structure and the existing craters are not just places of meteor and asteroid impact. Can there exist a civilization of creatures similar to people?

Patrick recalls Bławatska. It seems, however, that he does not know what Blavatsky said about the relationship between the Moon and the Earth. Bławatska thought that the Moon was the first in the solar system and "spawned Earth" when it was still semi-liquid and gaseous like some of the farthest planets of our system. When the Moon was suitable for settlement, Earth was still in the process of forming a globe. The new planet "sucked" water from the moon, because of its mass and gravity - centripetal force in the cooling process. I am not a specialist to comment, but many things suit me. Blavatska further said that on the moon the white race "Lunar Pitris" (the Moon Fathers) reincarnated. This information is in the letters she has translated, especially in the "Book of Dzyan". When the earth stabilized and became a favorable place for the development of humanity, a group of souls reincarnated in the form of very primitive Homo sapiens. After some time Lunar Pitris started arriving to Earth (Blavatsky does not write how whether they were reincarnated or flew in some vehicles), in order to ennoble and humanize the breed. Bławatska also writes about the fact that at some point Lunar Pitris began to lose the human element, and although their minds were more and more efficient and better and their bodies were still human, they ceased to be people sensu stricto. This was caused by the increase of ego and attempts to act on their own. Something like opposing divine plans.

Patrick is right about the terrestrial origin of UFOs. If there may be submarine bases (which we have evidence in the form of USO), underground bases may also exist. Certainly, there are (there are) bases on the Moon, but Earth is the "homeland" of these beings. Caves, underground, unexplored depth of the oceans. These are the only places where they can live without the fear of intervention and aggression of Earthlings. UFOs are therefore only "taxis" between both globes. Besides, because the beings formerly called Lunar Pitris still look like humans, a certain group of them is undoubtedly mixed into the human population. The name "Fathers" appears in all tribes on Earth and their gnosis. These Fathers are always white and have always been teachers of humanity and are up to this day. Aboriginal rock drawings clearly show them. I describe it in the "Secret of Hanging Rock". The cause of the last glaciation over the last half a million years was the "nuclear winter" after some conflict on Earth. However, this is a story much older than you do with Miloš Jesenským seems.

(This is a digression to our work: "The Mystery of the Moon Cave" and "Gods of Atomic Wars" - author's note)

15 - 12 thousand The BC was in the final stage of melting and the climate began to stabilize. It is still visible in Australia, which lost its connection with New Guinea and Tasmania, as well as in the configuration of the area, which has not yet been devastated by colonizers. On the rock drawings are human creatures in "suits" clearly contrasting with naked natives immortalized on the same paintings. Däniken and his followers see them coming from outer space, but to me these suits are more like clothes that protect against civilization from the underworld. These drawings, therefore, date visits for a period between 200,000. and 100,000 years ago, and their goal was probably to study how a small survivor of Earth's population bears elevated and increasingly weakening radiation. In some drawings there are objects of unknown origin, in no way resembling the usual Aboriginal tools. Visits are still a daily occurrence for Aborigines but their drawings no longer show people in suits. However, we certainly do not know, because according to government arrangements, most of the Aboriginal sacral sites have been closed to the white and protected by Elderly tribal councils. Only the Aborigines have an introduction and only to perform a ceremony.

Probably not accidentally Aborigines are from the beginning fierce opponents of uranium mining in Australia. Too fierce for "primitive" tribes, as white people see them. However, Aborigines still have contact with Protoplasts and Teachers who can inspire them. For the Aborigines, the places where the deposits occur are always taboo. They probably did not know exactly what was under the ground, but they certainly felt dangerous energy, the existence of which was confirmed by the Teachers.

Aboriginal gnosis claims that the Teachers were also their Protoplasts and Creator of animal forms and landscape. While the first two statements are quite understandable and explain the role of Lunar Pitris in shaping races on Earth, the third claim is much harder to grasp. It is known who is authorized to create - it seems that initially Lunar Pitris acted in accordance with the will of God, implementing his plan on Earth to create optimal conditions for the human races that they looked after. With time, however, as the Eastern magazines say, and behind them Bławatska, Lunar Pitris began to gravitate and deviate from the plan, acting on their own, playing the role of wizards. They made many mistakes by experimenting with animated forms, creating forms that were too big and / or multifunctional and too long-lived, which in the long run prevented them from feeding, while incarnate souls helped create too great a load of karma. In creating the landscape by using the forces of Nature as the will, they also made mistakes letting their imagination run wild and probably outdo each other in more and more crazy ideas and overusing paranormal forces.

So "glasnost" was made, followed by a thorough "perestroika" and Lunar Pitris lost the human element, or the possibility of returning to the source of their souls. Perhaps this moment has been preserved in many ancient cultures as the "rebellion of angels" and punishment, which was the consequence of the overstatement of their ambitions? Although outside science boast of its achievements and creates the appearance that it already knows everything about the Earth, it is admitted in a close circle that it does not know the basic issues and laws of Nature.

There are only suspicions, speculations, theories and hypotheses, some of which sometimes work, but most do not. It seems, however, that for some time, just from the Apollo program, certain circles on Earth have had a secret and decided to keep it at all costs. It is a conspiracy that has never been seen before, because the disclosure of "someone else" on this planet would undoubtedly destroy the order (or mess) diligently created by the last millennia. "Someone else", who has a technique far ahead of the leading countries of the Earth and whose true intentions nobody knows, because this "someone" has no intention of confessing to the leaders of humanity who are ruthless and deprived of humanity. Just the stronger one.

Although the caves and oceans themselves defend themselves against the curious, for some time the exploration of caves and oceans has stalled. Governments have cut funding for research, and what amateurs still do on a small scale is on the one hand, and on the other hand, it also faces various prohibitions. Such uprisings as the pyramids in Bosnia met with a tsunami of protests from the side of science, which should first throw itself at solving the mystery. Does not this mean anything? It is known that science is on the pay of governments manipulated on the one hand by business, and on the other by military circles. However, the pyramids have already become public property and it is difficult to break the case now. What archaeologists will find in Bosnia will show time. For now, it is suspected that in addition to the pyramids, there was a huge metropolis between them. There is of course the possibility that after the first significant discoveries the place will be fenced and sealed, as happened in other such cases. "They" probably do not want the world to know about them and put pressure on official factors in the same way as the Apollo program. Imagine how powerful they must be, since two proud superpowers fighting fiercely for the highest bid have suddenly fallen humbly.

* * *

Perhaps it was as Piotr Listkiewicz claims. We do not know. However, one fact seems to fit in, namely that the Moon is a world that cooled faster and its rocks evolved faster than rocks on the still cooling Earth.

And one more significant information in this matter, quoted from the Onet.pl portal:

In the Jagiellonian Library there may be a fragment of the first Tibetan Buddhist canon - "Kanduzuru" from 1606 - scientists assume. In the Jagiellonian Library, studies of Tibetan books from German collections are underway.

As Agnieszka Helman-Ważny, participating in the research, informed, the scientists are looking at a unique collection of over 800 books from the collection of the German scholar Eugen Pander , belonging to the so-called Berlinki. They assume that the collection may contain a fragment of the first edition of the Tibetan Buddhist canon - "Kanduzuru" - Wang-li edition from 1606.

"I found an article by a German scientist, Prof. Helmut Eimer, who suggested that one of the first editions of the printed "Kanduzuru", i.e the Buddhist scriptures, was lost during the war. The history of these collections suggested that what the Germans considered missing was probably located in the Jagiellonian Library," said Helman-Ważny.

As the researcher explained, research will probably last for many years. At the first stage, the researchers try to confirm above all whether the lost fragment of the oldest printed issue of "Kandżuru" is actually in the Jagiellonian Library.

"You have to review these books first and explain the fragments to find out what is in this collection. We know what the releases of 'Kanduzuru' look like and what they contain. Therefore, we know how voluminous it should be. Finding books with the text of Kanduzuru would not be such a difficult task. However, confirmation that this is the 'Kandżur' of Wangli's edition, and not one of the many later editions will require detailed research. The problem is also that the whole of these collections includes about 860 volumes, and the wanted Kandżur is only a small part of them," emphasized Helman-Ważny.

"Of course, at this stage we do not have exact translations or analyzes of each book, we just look for those that have the right number of cards, a characteristic format and which can potentially be 'Kandżurem'," she stipulated.

As Helman-Ważny explained, the collection is unique on a global scale. It contains manuscripts and prints mostly in Tibetan, but also in Mongolian and Chinese. Mainly they are religious texts, but not only. They are all very old.

"Our initial research already confirms that this collection was brought from Beijing to Berlin in 1889 by Professor Eugen Pander. Initially, they were sent to a museum in Berlin and later transferred to the Prussian Library. From there they came to the area of present Poland and, in time, to the Jagiellonian Library," emphasized the researcher.

As she added, it can be concluded from Pander's notes that part of the first edition of "Kanduzuru" went to Japan, and some were in those collections that were taken to Germany. It is known, however, that the volumes that got to Japan burned in a fire during the earthquake in Tokyo in 1923 and certainly no longer exist. "The only known parts of this issue that we know have leaked through Pander to Germany, and that's what we hope to find in Krakow," she said.

In addition to Helman-Ważny, the team dealing with the study of Tibetan books from Berlin is dealt with by prof. Marek Mejor from the Institute of Oriental Studies at the University of Warsaw and Dr. Thupten Kunga Chashab, also from the University of Warsaw.

Berlinka consists of two parts: the so-called Of the Prussian Treasury and a collection of new prints, i.e. books from the 19th and 20th centuries. The Prussian treasure is manuscripts, music manuscripts, old prints and a small fragment of cartographic collections.

The collections of the Prussian State Library were taken away by the Germans at the end of the Second World War, for fear of raids, from Berlin to the Cistercian abbey in Krzeszów at the then German Lower Silesia. After the war, these areas were included in the Polish state and the collections were taken over by the Polish authorities.

These collections from 1946 were transferred to "Jagiellonian" as the central Store of Secured Collections. 490 out of 505 chests went to Krakow. The owner of these collections is the State Treasury, and the Jagiellonian Library only their depositary.

The question should be asked: what did the Germans look for before and during the Second World War in Central Asia? Legendary Szamballi and her mysterious technology, through which was to create their magical Wunderwaffe? Something that had a premonition of Steven Spielberg creating the character and crazy adventures of Indiana Jones ? Or maybe they were looking for evidence that they were able to build and run a vehicle that would be able to move in time? Who knows if, instead of uranium and track ores, they did not look for lost artefacts in the Polish and Slovak mountains, such as the Moon's Shaft? This thought is not so stupid as if it looked at first glance. But more about this in the next materials.

One more thing about the Ural Tunnel described by one of the Russian authors is: The title of the original article reads: "From the northern Urals to the south, you can go underground!" This amazing interview was published by Yegor Usachev in the pages of the journal "Kalejdoskop NLO" No. 3/2008 dated 13 January 2008:

Question: From the north to the southern Urals underground! Do you mean it seriously?

Answer: Of course, exactly as you said. Only from time to time you have to go to the surface to go from one underground cave system to another. The first and obvious reaction to the words of the researcher from Jekatierinburg, Mr. Igor M. (he asked for anonymity) is laughter. The man who was under the spell of A. Wołkowa's fairy tale talked with a certainty to me. "Seven Kings of the Underground". But when the laughter ceased, I began to remember what is actually known about the caves in the Urals? And it turned out that the idea of Igor was not as stupid as she looked. The western slope of the Ural ridge turns out to be a perfect place for the occurrence of caves. Almost half of the mountains there consists of limestone, dolomite and gypsum fuzzy waters. In these stretched areas, there are areas with increased susceptibility to cracking and penetration of underground waters into the rocks. The old valleys of the Ural overlap with these zones. Over the hundreds of thousands of years, the process of karst formation took place on their banks and under rivers. Then the Ural mountains again came alive and began to rise, and glaciers appeared on the tops. When they melted, the younger and deeper valleys, often overlapping with the water network, grew into the former meridian valleys filled with post-glacial sediments. The cave systems, separated into separate fragments with earthen pitches, appeared above the groundwater level and became accessible to people. Such caves in the Urals were found above 500!

- For all ages, people tried to settle underground spaces for one or other reasons, continued Igor. They searched for valuable ores, built secret temples and shelters. Particularly many such artificial underground constructions are located in a triangle whose vertices are the cities of Perm, Jekatierinburg (former Sverdlovsk) and Chelyabinsk. The imposition of artificial underground structures on the existing cave system resulted in the creation of a huge underground system of corridors stretched for many, many kilometers.

P: Yes, but why has not anyone passed through this system to this day? We have many extreme athletes in the country who practice underground tourism. There are a dozen major sections of extreme speleology in Ural cities, are not they?

A: First of all, Ural caves have not been examined as thoroughly as the Crimean or Caucasian ones, to which speleologists from all over the former USSR were heading. In those cases when Ural expeditions were experienced speleologists with the habit of passing through clay barriers and dams separating cave systems into fragments, whole systems of unknown corridors and caverns were discovered. One of the caves, which turned out to be the entrance to the entire underground labyrinth, is called Diwija. It is located near the small city of Narym. The locals knew her for a long time but claimed that its length is only a few hundred meters. In the 1960s, a group of speleologists from the MGU decided to check the size of the Diwija cave. They investigated several corridors and dug up suspicious places. Their efforts crowned success. One of the narrow corridors led them to a whole, multi-kilometer cave maze with amazing beauty. But that's not all, because, in the words of the expedition's manager, behind the newly discovered part of Diwija, there is another huge cave system with huge rooms and galleries not discovered. The second reason may seem absurd, mystical, but it must not be omitted from consideration. I analyzed many unfortunate accidents that took place in the Ural caves, and I came to the conclusion that most of them are against the laws of physics. I have a feeling that something or someone in an unnatural way interferes with going through some caves. There are such limitations in them, after which one begins to feel a negative impact on the human psyche, something causes that it begins to behave illogically, commits dangerous acts, from which - to know why - cannot resist. Even very experienced speleologists make mistakes that they cannot explain later. They cannot, if everything ended well. And there have been tragedies, such as the multi-storey Sungan cave.

Q. What is this cave?

A.: It is one of the longest in the Urals, the length of sidewalks only on the lowest level reaches eight kilometers! It was explored by many speleologists and they all talked about auditory and visual hallucinations experienced by people in this cave. Speleologists repeatedly stormed this cave and, reaching the bottom of it, remembered the feeling of an incomprehensible, unjustified fear that caught them in one of the cave corridors. No one has passed this corridor until the present day. Others have again mentioned the unpredictable swing of moods during the exploration of this cave, which happened even to the most hardened of them. This circumstance may have been the cause of the tragedy that took place in this cave 40 years ago: two outstanding speleologists from Moscow died here - Valentin Alieksienski and Elena Alieksieeva.

The winter, multi-day cave exploration ended. The explorers climbed from the lower level and stopped next to the speleological ladder, made of special steel chains and aluminum steps (it was suspended on a special turnstile). They stopped by an amazing spectacle: the ladder covered with ice, because the entrance turned into a waterfall. Trying to climb such an icy ladder to 100% had to end tragically. It would seem that it was enough to wait a few hours until the water stops flowing or wait until someone from the neighboring farm approaches the turnstile and asks what the speleologists are. But Alieksienskij, dressed in a waterproof suit, went down the ladder, bouncing the ice off the steps and crouched to avoid the stream of water falling from above. At first, his companions saw him, and then they only heard him, and then he stopped talking to them. Then Lena Alieksiejew followed, and the same thing happened to her: at first she spoke to the cries, then disappeared from view and fell silent. Forever.

After a few hours, the local peasants who helped speleologists to get outside came up to the mouth of the cave. When the tractor managed to pull the ladder outside with the help of a tractor, a terrible sight appeared - Valentine and Lena turned into ice statues. Frozen to death!.

Q: So you convinced me that the caves punish people for curiosity. Well, but in the existing system of underground connections between the North and South Urals. Is there evidence to suggest such cave systems that consist of artificial and natural fragments?

Seeing that I doubt his words, Igor removed a folded sheet of paper from the shelf. Even the dilettante would have realized that two types of galleries were shown on the set. The geometrically correct network of artificially dug corridors is here, that is where the corridors and halls of natural origin are connected.

Before us there is a fragment of the underground system discovered in the vicinity of Ekaterinburg. You can enter it directly from the city, but this system has not yet been defeated by anyone. I have managed to pass through 48 hours, passing numerous underground streams, to the end of only one of these corridors.

Surprised? Well, let's open a book about the Urals, which was published by the USSR in 1979 in a limited edition. It turns out that when filling one of the former depressions between the towns of Niżnie Siernia and Njazepietrowskie, 23 out of 45 rock fissures found there had connections with underground spaces up to 10 meters high! The rock mass found there was almost half empty! So, who knows if going through these systems caves will not be able to pass from the northern Urals to the South?

Not only caves

Translating this text, I was reminded immediately of reading my youth "Journey to the Center of the Earth" by Juliusz Verne, "The

Secret of the Birds Blind" by Ludwig Souček and books by Maciej Kuczyński, Jacek Kolbuszewski, Vladimir Obruczew, Józef Sękowski, Erazm Majewski and other authors treating the mysterious spaces of the planetary underworld. I was reminded of the quote that Earth reminds an old house full of cracks and holes. In addition, every mention of cave systems still resembles a legend about Agharta given by prof. Antoni Ferdynand Ossendowski in his book "Through the land of people, animals and gods".

We have already mentioned previously Ural "pisanice" - rock drawings in the Urals. The geologist and mineralogist, Dr. Vladimir Tjurin-Avinij from Samara, is absolutely convinced that our planet has been visited in the past by representatives of extraterrestrial civilizations. In his opinion, these strange urinal rock drawings are nothing more than chemical imaginations! Every high school student knows this graphic way of showing chemical patterns, so-called "Structural formulas", used mainly in organic chemistry. The scholar created a bold hypothesis that these chemical formulas could one day be passed on to primitive people by Visitors from space, who once visited the Earth. In our opinion, the alternative to Visitors from space is representatives of the twilight civilizations inhabiting the Earth and described in ancient Hindu poems.

Let's recall the Russian Urals legends about people from the mountains who lived underground and identified them with gnomes, goblins and other Chthonic creatures, the Underground. Perhaps these are accounts of people left behind by the terrible conflict that swept away from the Earth's surface the highly developed civilization of the Empire of Atlantis, who for centuries inhabited the caves fearing to go to the surface devastated by the conflict of the Earth. It sounds as sensible as the hypothesis about visiting Visitors from space.

And to finish this chapter, we want to present one more hypothesis about the destruction of Atlantis, whose creators are Robert Leśniakiewicz and his sister Wiktoria, and here she is:

The Holocaust of Atlantis: Guilty of the cyclic "hot spot"?

The controversy over the existence of Atlantis lasts two and a half thousand years, i.e. since the publication of the account of this mysterious island. Some claim that the entire continent, by Plato (437-347 BC), one of the seven wise men of Antiquity, in two so-called dialogues "Timajos" and "Krytias", in which he described the Atlantean Empire and its extermination, which took place some 10-12 thousand years ago.

So far, no one has been able to give the true cause of the tragedy that took place around 10,000 BC somewhere in the Central Atlantic, two days west of the Pillars of Hercules, where the land of the Hesperides and Celtic Avalon lay. It is believed that the reason for the sinking of Atlantis were the powerful volcanic eruptions and this is what the encyclopedias say. Another theory involves the destruction of the Empire with the fall of a large meteorite or asteroid - a replay on a smaller scale event from 65 million years ago, when dinosaurs were destroyed. Either way, most hypotheses revolve around geophysical causes. So far, however, no one has connected this event with plate tectonics and its accompanying phenomena. We think that the existence of terrestrial and oceanic plates (so-called lithospheric circuits) and hot plumes located under them (also called hot spots or spots - in the world literature hot spots) may be the explanation for the destruction of this mysterious island, which had to exist on middle of the Atlantic. Another location is simply impossible, it would be probable.

As everyone knows now, the Earth consists of a dozen or so continental and oceanic plates that are constantly in motion. The continental plates are moved by oceanic plates, which are formed in rift valleys in the middle of the oceans and disappear in the subduction zones, which are oceanic trenches - the deepest places in the alliance. It is the subduction of oceanic plates that causes their motion and the motion of matter in convective cells, and thus the whole phenomenon called spreading . However, it was still not very clear, which in turn causes the generation of ocean plates, from which a constant inflow of magma to the rift valleys and volcanoes. The answer came again from the depths of the Earth: from the inner layers of our planet, powerful plumes of hot matter beat their shells and break through to the surface creating "hot spots". So far, scientists have found 40 such places on the surface of the Earth, where the magma flowing directly from them escapes to the surface creating powerful volcanoes. These include Galapagos Islands, Azores, Cape Verde Islands and Hawaii where there is the most powerful mountain on Earth measuring over 10,000 m from the bottom of the ocean to its apex and Iceland - which is an elevated fragment of the seabed and one large volcano.

Iceland. It seems to us that it is on this island of mists, winds, volcanoes and glaciers that there is a key to understanding the mystery of the destruction of Atlantis. As it has already been said, Iceland is a fragment of the sea elevated above the ocean level. The main question is: what caused this elevation? After all, Iceland is cut from Vík to Raufarhöfn by a rift going through Vatnajökull, along which magma gets to the surface of the Earth and thanks to which the island exists on the map of the world, on the northern part of the underwater ridge of Reykjanes. This is true but not yet whole, because under Iceland there is a hot plume, whose upward pressure causes the elevation of this fragment of the ocean floor to the height of almost 4 km above the level of the ocean floor. Evidence of this is the ongoing volcanism on this island, which is geologically young and has no more than 65 million years. Under its largest glacier, Vatnajökull, there is the Grimsvatn (Grimsvötn) volcano, which is about 4-10 years old -

1,719 m above sea level and Hvannadalshnúkur - 2,119 m (the highest peak of Iceland), and 138 other volcanoes, 26 of which are constantly active.

What are we heading for?

We are heading to the fact that Atlantis was perhaps such an elevated area of the seabed in the reservoir lying directly in front of the Strait of Gibraltar. The island (or even the archipelago) was large enough to block the flow of the Gulf Stream, and thanks to that, ice sheets appeared and disappeared from time to time in Europe and North America. Atlantis may have been inhabited by some civilization that may have had its colonies in Europe: Britain, Brittany, Spain and part of the North African coast, and in the Americas, especially in the Caribbean basin. In this case, Sir Brinsley le Poer-Trench - Lord of Clancarty, who in England sees the province of Atlantis is wrong. (=> Brinsley le Poer-Trench - "Men Among Mankind", London - New York 1962 and "Operation Earth", London 1976 - on the website of the System of Love of Nations - www.sm.fki.pl and WWW.hyboriana.blogspot. com) It was only her colony and nothing more. While we were in Portugal, Ceuta and Morocco, we found a lot of small stones in red, white and black on the local beaches. To this large number of transparent pebbles, as if from bottle glass, which could not be the remains of meteorites or tektites, whose nearest field is in the Sahara. The colors and houses of Atlantis had such colors. Bottled glass, which we think was of volcanic origin, was also found by our friends from Florida, and colorful pebbles too. Is this just random coincidence? There is no such "accidental convergence" - Atlantis really existed! Atlantis existed thanks to the plume of hot magma, which elevated it to a height of at least 4,000 m above the level of the Serenity.

What happened 120 centuries ago? We can only assume. Colonel James Churchward - the creator of the Pacifida theory or the Mu (Moo) continent, thinks that Pacyfida stood on huge underground void, which collapsed under the influence of giant terremoto. There is something in it, but underground caverns should be substituted with a "hot spot" that pushed the ocean floor upward, creating a huge bubble on the surface of the ocean floor. Imagine now that a small asteroid is in this bubble. What will happen then?

The energy of the impact will be absorbed by the liquid matter of the plume, but the bubble itself, like a pierced balloon, will fall to the bottom of the ocean, burying what was on its surface under the waves. And one day and one night is terrible - wrote Plato. This would explain the relatively small losses suffered by the nature of our planet from this impact 12,000 years ago and the fact of survival by humans and smaller animals of powerful earthquakes and tsunamis and the fall of the Gulf Stream into the North Atlantic and the resulting melting of ice bowls in the northern hemisphere of the Earth. But the above explains only one episode - the youngest, 12 million before. Additionally, there were four of these glaciations and interglacials!

Perhaps it was different - the hot spot is just being created there from the central Tertiary and its "work" is still unstable. The bubble created by its activity already appears above the surface of the Atlantic, it is already immersed in its waters, which causes cyclic glaciers and interglacials. So it looks like the island of Atlantis has already lived through four cycles of its "life".–I If you believe Edgar Cayce and other visionaries and messages that some contactees receive during telepathic contacts with Aliens, soon its fifth appearance will occur on and in Atlantic waters.

The "hot plume", which is responsible for it, is still active, only that from 12,000 years its activity has decreased to a minimum and the ocean floor is located at a depth of 6,000 m below the surface of the Atlantic. Volcanoes are open in the Canary Islands, which is the only tangible proof of the existence of this island. Submarine volcanism in this region of the Atlantic Ocean also concentrates on the Azores, Madeira and the Canary Islands. Recent sonar research of the Atlantic bottom in this area also shows a seabed topography similar to the topography of Atlantis from Plato's descriptions.

What does it mean? Well, there was not a violent cataclysm and the island of Atlantis went down relatively quietly, otherwise the mountains would be rubbed off with stones. The first signal of the upcoming change was the calming down of the volcanoes. Of course, everything took place calmly on the scale of the planet, because for people it was a terrible cataclysm - one day and one night terrible.– However, most of them managed to save themselves and escape to the colony. We know the rest.

Was it like that? We do not know for sure, but there are many indications. If this is the case, we will see it soon, when the "hot plume" will increase its activity and the ocean floor will start to bear in the Azores area, and the volcanoes there will start violent eruptions. Of course, the birth (re) of a new island the size of Iceland (103,000 km^2), or even greater, will be a violation of the ecological and tectonic balance of our planet, which will also affect us in the form of a new glaciation of Europe and North America. Perhaps it will ease the greenhouse effect a bit, but maybe there will finally be decent winters in Poland?

And one more significant thing - as it seems to us - insight, namely - look at the reader on the shape of the bottom of the Atlantic between the Azores and Gibraltar. There are underwater mountains standing on faults. Now look at the western side of the rift. On the other side, there should be a similar formation, but there isn't. It is a rule that every formation on the eastern side of the rift has its counterpart on its west side and vice-versa, for example the equivalent of the Cape Verde Islands are Bermuda. In this case, this rule has been violated! The scholars trying to explain this anomaly somehow introduced an extension of the separation between the stationary African Plate and the Eurasian Plate pressing against it from the North-West, but it certainly did not create an underwater elevation. This reminds us a bit of the situation prevailing in the so-called Afaru's Tri-node, NB, under which there is also a "hot plume". As you can see, the hypothesis of a hot spot under that area for the Atlantic is gaining a strong point.

We will live and see - the worst is always ahead of us - which is adequate because we do not have any influence on geological changes and the only thing we can do is adapt to them. So let's wait and let us be ready.

CHAPTER VII

Megaliths - energy aspect

From the beginning of the blog, the Research Center for Anomalous Phenomena - WWW.wszechocean.blogspot.com - currently under the name Wszechocean and the auspices of the KKK, there are articles by Zofia Piepiórka - Eleonora, which has been researching them for many years.

There are also a few of our concepts on the blog about their origin. In principle, we are convinced, similarly to foreign researchers, that they are a kind of echo of the splendor of ancient civilizations - the civilization of Atlantis, or even older Atlantis civilization. Personally, we are convinced that all this knowledge has already been known to people and has been forgotten.Now let's look at it from the point of view of these people. Who were the people who had such opportunities? These people were the gods of our planet!

Recently, Robert Leśniakiewicz argued with the opponent of the theory of the existence of Atlantis, Interterry or Atlantic, who claimed that if these civilizations existed, traces of mining works would remain after them. Yes, this man was right but, first of all, if these mining works were carried out, traces of them would be obliterated by time - after the collapse of the Atlantic Empire 12,000 or even 20,000 years passed! * Second, even if some machines left in the rock mass, after that time they would turn into nested iron ore deposits with additions of non-ferrous metals and perhaps REE[18]. There is still an alternative to this, but more on that later. And thirdly, a civilization that could create one of the elements from the other on an industrial scale, would not have to play with ore mining, able to carry out the transmigration as needed.

After the fall of the Atlantean Empire, which could have happened as a result of civil war, war between civilizations or ecological catastrophe, Atlantis' knowledge was passed from father to son until it lost its original meaning. All those people I mentioned at the beginning have partially discovered the sense of the Atlantean messages and perhaps they have succeeded in extending life by years, which of course aroused the vigilance of the Holy Inquisition with known consequences: torture, trial, stack or life imprisonment and confiscation of property for the Church. After all, the princes of the church also wanted to live for a long time, and that with the help of what they called witchcraft, and in fact it was only distorted, degenerated ancient knowledge.

During the VII UFO Forum in Wroclaw, in April 2003, we had the opportunity to see what the team "I cannot believe" filmed in India. These were the performances of various fakirs, which reminded me of the attempts that astronauts make before they fly to space. This is also a cultural message from Atlantis. Let's not forget that Hindu culture of 8,000 years is one of the oldest cultures on Earth! It is in the "Mahabharata" and "Ramayana" that we find descriptions of flying

---

[18] Rare earth elements whose atomic nuclei are unstable resulting in radioactive.

machines and other devices, including weapons that existed even earlier than these poems were written. It is in these poems that people are written about flying into space and settling other planets. It is in these poems that we write about ancient civilizations that have a huge - even people of the 21st century know us! Exactly the same message is also alchemy - only from another field - chemistry and biology. Let's just think - creating a homunculus is nothing but creating a transgenic human. A new creature with a mutated genotype. And all those known to us from Greek, Chinese, Japanese, Hindu, Scandinavian, Slavic, Native American, Negro legends, legends or myths monsters, chimeras, cyclops, dragons, werewolves, etc. hybrids of living beings, so what? These are genetic engineering products! Perhaps they were simply biological machines, robots that served people and worked for them.

Such a biological civilization would be deprived of all the "favors" of our (supposedly magnificent) civilization: dirt, stench, oil spills in the oceans, radioactive poison, millions of tons of waste and rubbish! And dying would not leave any traces - except those which were built of the most durable material: stone!

But let's go back to the megaliths. These words were written in 2004, and so far the advances in genetic engineering indicate that we were right. GMOs[19] are a fact that we see and even consume daily. But what was described was not a vision, only a strictly scientific prediction based on facts, not esoteric ideas.

Now let's remind you what the ATLANTID is about. Legendary land mentioned for the first time by Aristocles the Athenian aka Plato. He said:

Atlantis lay behind the pillars of Hercules. It was a great power, its area reached up to western Europe and Africa.

---

[19] Genetically Modified Organisms.

According to Plato, it existed about 9 500 years ago. Atlantis sank, covered by the ocean's waters. It has been debated for a long time whether this was just a myth created by Plato, or indeed this land existed. Already at that time, many of Plato's disciples were arguing with each other.

Francis Bacon in 1626 also mentioned this continent. Legends, myths and stories about Atlantis and its extermination in the twentieth century have absorbed people more and more.

Scientists point to many places that may be the forgotten continent. The islands in the Atlantic Ocean, including the Arctic, Azores, Bahamas, Cuba, the Caribbean and Bolivia, and in the east, Crete, Santorini, North Africa and the Red Sea are considered to be the area that occupied Atlantis. Scientists and various researchers of Atlantis describe its culture and great power.

This was:

- the dominant power in the world about 10,000 years ago, and

- this land is still very secret to the world.

The biblical flood in the Old Testament and 700 years of Noah's life are identified with Atlantis.

After seven days, the water of the flood fell to the ground. In the six hundredth year of Noah's life, in the second month, on the seventeenth day of this month, on that day the springs of the great abyss gushed out and the rebates of heaven opened. (Gen. 7: 10-11).

In his Atlantean dialogue, Plato presented 11 descriptions and theories about the culture of Atlantis. From his dialogue, it appears that this continent was larger than the entire Arabian peninsula, and God Poseidon ruled him. The name of Atlantis derives from the first son of Poseidon, Atlas. The land located behind the pillars of Hercules, beyond Gibraltar, was divided into ten different territories on the Atlantic Ocean. Plato described Cerne the capital of Atlantis as a big city, there lived a very rich society. They dealt with politics and had an army. Egyptian records describe the wars between the Athenians and the Atlanteans, they also mention that both these civilizations were destroyed during the great catastrophe that occurred in those days.

There are many reasons mentioned that could have destroyed the culture of the Atlanteans:

1. Her own power

2. Striking a large asteroid or comet

3. An earthquake, great volcanic activity and rising water wave

4. The rising water level caused the melting of the glacier

5. It also mentions the polarity of the Earth and drastic weather changes caused by this event

However, all cultures on Earth talk about the great flood in this region. The ancient Greeks supposed that it happened about 10,000 years ago. Plato heard the story of Atlantis from Solon, besides him they also talked about Atlantis Edgar Cayce, HP Bławack, this story is not foreign to the Hindu Brahmins and Druids. In the book "The story of Atlantis" by her author Wiliam Scott-Elliot, an anthropologist gives accurate descriptions of Atlantis, he points to the peoples of Mongols, Semites and Toltecs who can be descendants of Atlantis. technology was alien to them, but as they carried the news, they used their powers to control and manipulate other people in a diabolical manner, their moral fall was the result of their own fall.

The geographical location of Atlantis on the globe is differently placed:

- in the Atlantic,

- in the Mediterranean Sea,

- in Western Europe and in Africa,

- the Caspian and Black Seas are also mentioned,

- there are those that put this land in the Gulf of Mexico

- and in the basin of the Mississippi River.

One of the territories of the destruction of Atlantis is the destruction by hitting an asteroid or comet in this land, this view was presented by Ignatius Donnelly. The great impact caused by this blow dissolved the glacier and the land sank. Edgar Cayce often said at the time of his prophecy that Atlantis sank. He claimed that it was an island the size of Europe and Russia. He located it in the area from the Mediterranean Sea to the Gulf of Mexico, according to him extended over most of today's Atlantic Ocean. Cayce also believed that the ruins of the temple found in Bimini near Florida are a remnant of Atlantis. He described the Atlanteans, experts in advanced technology. Plato claimed that the Atlanteans were more developed at sea, had excellent ships, but also engaged in gardening. Cayce also said that Atlantis already existed about 50,000 years ago and it disappeared about 10,500 years ago.

About 28,500 years ago, this continent was taken over by three other islands: the Arians, Poseidon and Og, and their inhabitants led to the destruction of Atlantis. Those who managed to survive went to other lands on Earth.

A new hypothesis by Wiaczesław Kudriawcewa from the Institute of Prehistory, based on the Plato dialogue: Great Britain, Ireland, northern France are a remnant of Atlantis.

There are many other theories regarding this land. Scientists are looking for Atlantis in various places on Earth. There are also those who express opinions that this land is lost a little later, about 8 500 years ago.

There is always great interest around Atlantis, there is a lot of discussion and even more controversy. Many scientists and archaeologists are studying Minoan culture on the island of Crete and the island of Santorini. The ancient Minoans were powerful, they dominated the Cyclades.

The island of Santorini lies near Crete, destroyed by the volcano 1,600 years ago. Geologists say the first massive volcanic eruptions in Santorini took place 23,000 years ago. The last very devastating, disrupted the area of all of Europe, Asia and Africa 1,600 years ago.

There are a lot of mystical and mysterious lands on Atlantis attributed to Atlantis. After the total sinking of Atlantis, a large group of its inhabitants emigrated to Egypt. The great priests called Ra Ta gave rise to Egyptian culture. Atlantis saw the most secrets of the world in 1958-1998. Old theories preached by Plato, HP Błavacka, Rudolf Steiner, Aleister Crowley and Francis Bacon were confirmed.

In 1970, Dr. Ray Brown, near the Bahamas, discovered stone structures resembling a pyramid under water. During the diving and research of these great underwater monuments, a source of light was noticed. Divers followed in that direction and saw a four-inch crystal ornamented around red stones. Brown said that this crystal had great mystical power. People in his vicinity felt a strange wave, like a gust of wind. . Light was observed and voices heard and everyone felt strange vibrations in the water surrounding them.

The Bermuda Triangle in the Atlantic Ocean is a huge mystery to the whole world. It stretches between southern Florida to Bermuda and Puerto Rico. Area called the Bermuda Triangle, famous for many mysterious and mystical events. Vessels disappear and planes die. Many theories try to explain this strange phenomenon, scientifically unknown. So far, to no avail. The Bermuda Triangle, also known as the Devil's Triangle, absorbed many lives in a mysterious way.

Edgar Cayce (1877-1945) left a lot of information about Atlantis. One of his prophecies says:

Part of the temple can be discovered under sea water near Bimini. I expect this not so long around 1968 or 1969.

Twenty-three years after the death of Cayce Bimini, road was discovered about 15 feet (5 m) underwater in the Atlantic Ocean near the island of Bimini on the islands of the Bahamas. Bimini Road is a mile long, made of beautiful stones, especially the beautiful view from the plane. People have always thought that these rocks underwater are a work of nature, but now they have some doubts about it. Mystical rocks have been found under the water, they are not accidental. One of Dr. Little's researchers does not deny that it may be Atlantis and these rocks resemble an authentic port. Similar places were found in Cuba, Asia and Africa. Regardless of the statements of dr. Little of several researchers today believe that Bimini Road is part of Atlantis.

On the other side of the Atlantic Ocean, scientists are investigating the island of Spartel near Gibraltar, it is now 60 meters underwater near the place that Plato described:

There is an island in front of the pillars of Hercules, the island is as long as Libya and Asia together. It was the way to the second island and from them it was possible to enter the continent.

Geologists have determined that the place of the Spartel Island was destroyed by earthquakes and tsunamis about 11,600 years ago. Plato claimed that it ceased to exist 9,600 years ago.

One of the most famous legends describing Atlantis comes from Solona, he gave it to Plato while he visited Egypt. The famous dialogue "Timajos" and "Kritias" acquaints us with a previously unknown land, called Atlantis. It is a story about the great civilization of the Atlanteans and its destruction, caused by a volcanic eruption and a powerful wave of tsunamis.

Scientists suppose that the island of Santorini may be the remnant of Atlantis. At that time, the Minoan civilization in Crete also died. Archaeologists have discovered old ruins in Crete and Santorini. They are from the same time. They also come up with the theories that its inhabitants could have been Atlantis.

The great event was found by archeologists Akrotiri, near the capital of Santorini Thiry (Fire). The people who lived there resemble their Cappen culture Minos culture. Finding Akrotiri was a big event in the archaeological world.

The volcano eruption in Santorini has had a great impact on the weather all over the world. Volcano eruption destroyed the entire island. The volcanic crater is located on the island of Kaimeni - a fragment of the old island of Santorini. 1600 years ago the largest volcano in the world ripped the Island of Santorini to shreds. Today's Santorini consists of five fragments - five small islands. The main and the largest island called Thera, about 70 km long and very narrow and the second smaller Therasia are inhabited by people. The other three islands are only volcanic rocks. On the island of Kaimeni, the ground is hot underfoot. Almost every day gas vapors emerge from the crevices of the crater. The volcano on Santorini is still dangerous. In modern times, the Kaimeni volcanic crater ejected live fire eleven times. The last eruption was in 1950

Plato in his dialogue "Timajos" and "Kritias" left information about puzzling land known as Atlantis for the world.

He had heard the old legend from Egypt from Solona and he had met the priest from Thebes Sonches before, who read the hieroglyphs on the columns of the temple, describing the conflict between the Atlanteans and the Athenians.

Plato accurately described the capital of Atlantis. It was amazing in terms of architecture and engineering. A few round rings lying on the water were created from the mainland. They looked like an eye. In the middle of the temple was built Poseidon - God of the sea.

The golden statue of Poseidon racing on a chariot, which was led by six winged horses, was the symbol of Atlantis and the adornment of this place. The Atlanteans around the capital built a great system of canals and bridges connecting all the rings of land, secured passages with gates decorated with bronze and built stone walls.

According to Plato, Atlantis was destroyed 9,000 years ago, so the gods decided. They sank this land during one night.

The sleepy prophet E. Cayce has placed Atlantis closer to the Bermudian islands. He believed that they had high technology, including powerful "fire crystals" with extraordinary energy. The fire crystal was caught in the wrong hands, got out of control. This Cayce situation caused the sinking of Atlantis. The energies of this crystal had a nuclear power. As well as in Cayce's prophecy, it is said that a very strong ocean wave was the cause of the tsunami, Cayce suppose that the destruction of Atlantis caused great energy.

A lot of scientists pay attention to the culture of Crete and Santorini. They are looking for a connection with Atlantis. They also wrote about the lost continent: Plutarch, Herodus, Timagenus, Theopompos, Greek historian, J. Churchward, author of many books about the Earth MU and Dr. Gorge Hunt Williamson, author of books on Atlantis and Lemuria - researcher of these lands and anthropologist. In his opinion, both the Atlantis and Lemuria lands were destroyed by the powerful Atlantean technology. He found sources in Peru, in the Andes Mountains in the Inca temple, and kept old records.

Hopi Indians express similar views.

Disasters by fire and water for landowners are not new. Different regions have already experienced similar hits in modern times. So scientists do not exclude that Atlantis could die in a natural way. Many scientists also mention the glacier, could have contributed to the destruction of Atlantis Most researchers say that the Atlanteans were not spiritually developed, they adhered to the cult of Zeus. In their opinion, it is for this reason that the Gods destroyed this land. God Zeus led many wars, he is credited with unleashing the Trojan

War. He fought with Gods, he says mythology decided to destroy a man from the Bronze Age. He brought big rain and flood to the world. Only a small handful of people survived, and Atlantis also died. For Zeus, human bodies were sacrificed on its altars, especially young boys. After these ritual rites, its participants could turn into a wolf.

God Poseidon dominated the sea, Zeus in Heaven. They were brothers.

The father of Poseidon and Zeus was Kronos.

When Zeus grew up he protested against his father and won.

He also fought with the rulers of the Earth.

Solon gave Plato, who was born 600 years after the Trojan War a legend of a great flood and told the story of the sunken land and its inhabitants. The Egyptians had great knowledge, they knew periodic destruction of the Earth by fire and water, as well as by lasers, which have already destroyed humanity more than once. Those who lived inland were suffering more than residents near the rivers and seas. Mountain residents could survive. Big cities by the sea were picked up by the water. Only people who could sail could cross the safe land. The Egyptians possessed many records from previous eras on earth, and the memory of the deluge of Atlantis has also been preserved. Solon claimed that the Athenians had existed 8,000 years earlier than the Trojan War in 1200 BC

The same sources show that the Athenians invaded the Atlanteans, residents on the island bigger than Libya and Asia Minor, which lay behind the pillars of Hercules. Atlantis at that time was ruled by the king. He held in his hands still other islands and their rulers, he had great power.

The Egyptians said there was a time when the Atlanteans occupied the area of Libya, Egypt and the south-east of Europe. The Lord of Atlantis also wanted to subdue the Athenians. However, they had a strong army. Their first king was Actaeus, long before the deluge. His daughter Aglaurus married Kekrops, thus he became the king of Athens. King Kekrops is associated with the legend that his body was made of the body of a man and a serpent - this snake was called the "son of the earth," sometimes he was called "son of Gaia" (or Gei). At that time, each city had its own patron - god. Athens had two gods: Poseidon and Athena, who gave life to the olive tree and from that moment became the patron of Athens.

Luminous Athena, known as the goddess of wisdom, justice and courage, won the victory with all who were not righteous. She also taught people how to grow land, crafts and other arts. To all those who lived according to the law and enriched their lives with real fruit, he rewarded abundantly and blessed their homes. She was the first founder of the court, where murderers' trials took place. She was against killing and doing evil. We owe Atena that murder was no longer passed with impunity. You could not kill a mother, father, husband, wife, child or other human being.

Words: there will be no honor until death for a woman who will kill her husband. They are attributed to Atena.

Returning to the King of the Kekrops, some believe that he defied Zeus. He refused to sacrifice blood, he did not want to kill any life on the altar of Zeus. Kekrops was a powerful ruler. When the Gods divided the Earth into separate regions, the island of Atlantis came into possession of Poseidon. He settled there, married a mortal woman, Kleito, and had children with her.

The middle part of the island was a plain but in the middle there was a mountain. It was there on the hill that Poseidon built an altar, around which he carved round belts divided by water in the ground. In this way, he strengthened the entire island and its circulation. He made her warm in the spring and in cold winter, so that the earth could produce good fruit.

Poseidon had five pairs of twins, hence the division of all of Atlantis into ten provinces. His land was named Atlantis from the firstborn son of Atlas, who reigned over his siblings. His brothers were rulers in their royal palaces, they also managed the territories and people around the Mediterranean. Their kingdoms reached Egypt and Tiscani (Italia)[20].

Ten kings ruled in Atlantis, each in his province. The inhabitants of Atlantis were rich, they had animals, even elephants, they developed gardening, rich in fruits and vegetables. They built temples and expanded their country. They had led many times to war with the Athenians. But as legends carry, these people could not be happy because they did not develop properly in spirit. Their great obstacle was Zeus and the gods, worshipers of blood cult. They made blood sacrifices on their altars. Many Atlanteans accepted their rite.

The largest island of Atlantis was called Poseidonia (the island of Poseidon) and there were two smaller, very beautiful ones:

The power came from the sea from the Atlantic Ocean. The capital of Atlantis was situated on the water on the rings of the earth. On the central hill was the temple of Poseidon with a large golden statue of Poseidon, who was driving a chariot drawn by winged horses. Atlantis kept the law and its rulers ruled justly, worshiped Poseidon. Everyone lived a simple life and had beautiful qualities. But changes were slowly taking place. Zeus saw the greatness and immortality of the Atlanteans, rebelled against other gods to punish them and destroy them (Plato).

At that time, one island was situated opposite the pillars of Hercules. This island was larger than Libya and Asia. There were also other islands, from which one could go to the continent, which was

[20] It is probably about Germany (Tyskland, Twiskland, Tedeschi), not Italy.

surrounded by the ocean. A wonderful superpower ruled over all islands and other parts of the continent.

And quoting the words of the legend, we learn:

It was a big island like Libya and Asia, but after the great earthquake it sank. Covered with mud, it lies at the bottom of the ocean and is inaccessible to the eye of sailors. (Plato)

The annual return of birds from Europe to Central America keeps large groups of birds in the Atlantic Ocean in one place. Scientists assume that these birds still have land memory. They always rested on it when they made their journey.

Atlas often stormed Heaven, he opposed Zeus. And he punished him and put all the earth on his shoulder. Following the Greek mythology, god Atlas was a very clever god. Karl Kerenyi - the mythologist calls him the first astrologer, who held in his hands "dangerous wisdom", knowledge from the stars. He supported the pillars and separated the sky from the earth. The wars between Atlas and Zeus lasted ten years. The earth has many sources of energy, the sun is the biggest. The Earth itself also produces large resources. Built in a specific way, according to secret geometry, like the human body is connected by meridian systems and points, it abounds in a large number of huge energy reservoirs (earthly chakras) and energetically stronger places and channels through which all these places are connected with each other.

This is not modern knowledge, Plato (427 - 347 BC) recognized the Earth's grid. He preached the theory that the geometric structure of the Earth is responsible for the earthly elements and forms of life. Earth, fire, air and water - the four most important elements are inexhaustible sources of life. When they are balanced, they give off the perfect energy. According to Plato, the Earth's grid is made up of triangles, squares, cubes, etc. These geometric forms give great power, they generate, that is, the energy of shapes. So that's why it is so important that man does not disturb nature, he does not destroy nature, he did not change the great laws that govern the earth eternally.

People built pyramids on Earth. The Great Pyramid of Giza is best known to us. Scientists know that it is related to the Earth's energy grid and the blue bodies in the sky. The Atlanteans built many pyramids that rest today at the bottom of the Atlantic. There are those who believe that their sinking caused the disconnection of human consciousness from the Great Source. The great transmission of cosmic energy to Earth has been interrupted. The human mind has become unconscious and since then man lives in the dark, he does not understand the sources of his own origin.

The Atlanteans (or Atlantis) were known for their great knowledge of crystal energy. Crystals have great power. Whoever can use them properly releases the powerful energy that lies inside them. But only its proper application brings great values. When a man uses it inappropriately, he can be destructive in his actions.

THE RIGHT APPLICATION OF CRYSTALS

- treatment of various diseases,

- meditation - used during meditation give spiritual awakening, strengthen the paranormal abilities of a person,

- increase the capacity of the mind and the purity of thinking,

- they serve science and technology,

- with their help you can dematerialize, you can teleport,

- they strengthen the energy field,

- they are like a library - a warehouse of memory and knowledge,

- they are used in the development of plants,

- they control the weather,

- large crystals can be an energy source for big cities,

- they have the ability to transfer energy, are able to properly manage it, are an excellent relay,

- large crystals - with powerful powers, called "fire crystals" transfer the energy of radio waves between stations, cities, buildings, continents,

- crystals are also a power at a higher spiritual level, help to leave the body and move into another dimension, sometimes forever.

We know that the Atlanteans were famous for their knowledge and developed their own technology based on crystal energy. He often mentioned this mystical instrument in the hands of the Atlanteans of E.Cayce. He described the structure of the crystals used by them, their shapes, and described their energy, according to him the Atlanteans were able to connect with them the source of Earth's energy with the powerful energy of the sun, the moon and other stars. They made various precious, cylindrical and pyramid shapes of various sizes. They placed on their tops, different reinforcements, in this way they more closely focused the incoming rays of energy, pulling it from the celestial bodies and sending it to the right places.

They used this energy for various purposes. At the beginning, about 50,000 years ago (according to some sources) only for the purposes of their own spiritual development. Crystals were a tool by which man grew in a positive way. The early Atlanteans were very spiritual people. They developed their physical-material bodies for the needs of the spirit, used crystals to rejuvenate them. We read in the Bible that in old times man lived a very long time, hundreds of years and constantly looked young. We wonder why today, despite the great development of medicine and technology, human life is short. The ancient man knew how to use the powerful energy of crystals to strengthen his own body and prolong his life.

Somewhat later, the powers of the crystals were directed for other purposes. Their energy was transmitted from land to land like radio waves, ships and spacecraft were strengthened.

There are also such opinions that the Atlanteans could transmit long-distance human voice (phone?) And even images (today's TV?).

Fire crystal – Tuaoi Stone

The most powerful and mystical crystal mentioned by Cayce was a fire crystal - a stone called Tuaoi.

A large cylindrical prism, with six walls. At its top there was a special reinforcement that focused the incoming rays of energy and directed it precisely to the intended place. The Tuaoi stone was connected with the energy of the sun, the moon, with the earth's energy and with other unknown elements.

It is mentioned in the Bible:

Over the heads of living beings there was something in the shape of a vault that shone like an amazing crystal stretched overhead. (Ezekiel 1:22)

Probably the fire crystal fulfilled the role of a generator sending energy waves. Albert Einstein studied the predictions of Cayce and speculated how to use a 1 meter crystal quadrangle to achieve a similar effect.

There are three places on Earth that probably had a record and knew the design of a fire crystal.

1. Atlantis - Temple of Poseidon, near the southern Azores.

2. Temple of Poseidon, near Bimini

3. Egypt and Yucatan.

The energy of this crystal is compared to a laser, it was a powerful stream of energy with different colors. Each of these colors had a different wave. Cayce believed that the Atlanteans used the crystal fire energy to connect with the Supreme Source.

In the world of Atlante culture researchers, it is said that similar sources of energy are found in the Great Pyramid of Cheops and the Great Inca Temple, called Inkalithlon.

The energy sent by the fire crystal - Tuaoi is not known to today's science. It arouses great interest of NASA. Until Tuaoi stone was used properly served Atlantom for great purposes:

- they neutralized negative energies with it,

- healed the aura and the chakra system,

- improve the vitality of people, and

- activated human contacts with the Universe.

But when they increased their energy even more by their own stupidity, its high vibration caused the volcanic activity of many places and eventually led to the destruction of all of Atlantis. The great power of Tuaoi acted like a nuclear bomb. In addition to crystals, they used other energies, e.g. precious metals. They knew perfectly the energy of shapes - the law of secret geometry. The most known by us is the Atlantean ring found in the Valley of the Kings in Egypt. Its geometrical pattern balances the environment but it is important to be able to use it. There are different opinions about it. Some believe that they have energies that strengthen paranormal abilities and spiritual power, that they bring happiness, heals diseases. Others consider him a source of bad energy and blame for various misfortunes. But this ring, before someone starts to wear, should first recognize its harmonious tones. The Atlantean Ring is equated with the energy of the spiral - one of the first secret religious symbols that are the language of God to people. Spiral Consciousness (snail) marked Phi (φ), the so-called "golden value" (philosophy) is used in astrology and mathematics.

The golden value - φ = 1, 6180339887. This value has been known to the world for at least 2,400 years. Pythagoras - the mathematician gave us his definition and presented the value of φ more closely.

According to mystics, the spiral is presented: the circle of life - the time of karma - the time of reincarnation and human evolution.

The Atlas Ring, if properly recognized, evokes consciousness from higher dimensions, strengthens positive vibrations.

Atlantean crystals developed their powers in various environments. Their shapes were important, they gave them more and more new forms, from simple geometrical figures to more developed flower forms.

The flower of life - a crystal form - known by us as the Star of David - Merkaba was known in Atlantis. They were able to put beautiful geometrical forms out of the smallest crystals and program them for their own purposes, some forms for useful purposes and others for destructive ones.

Also, they were used as legends and prophecies for wrong purposes. They disrupted the natural sources of the Divine Energy. Their misuse has led to the separation of the body and soul, the distance from God. Man began to grow in material strength, his slogan became - "I will be a god." He forgot that he is one, one thread of the Great Consciousness. This great isolation has closed the great gates for man. It does not go unpunished when man uses powerful powers for his own selfish goals. Material building of life will not allow a man to dress a crown (lotus 1000 plane). Cosmic sources must be in complete harmony, then they will develop wisdom and open the gates to eternity.

Disturbed energies are a source of human destructive, earth and cosmos.

They are so-called "rays of death." (Or maybe it's just a negative information for the living matter that interferes with the proper functioning of cells and their aggregates - tissues that are specific rays of death?) Cayce said that the priests in the temples of Atlantis placed large crystals on their very front in such a way that when they stood on the altar the energies of these crystals could strengthen their power and connect them with God.

In Egypt, descendants of the Atlanteans also built pyramids and temples, used solar rays strengthened with crystals in order to heal people.

Crystals have great powers, their great source of energy can be used to treat cancer, to rejuvenate the body, unblock the energy points and the human chakras. A properly directed crystal energy beam gives the power like a laser. It can be used for spiritual operations in the human body. If a man learns to use this energy properly, it will be a great success and a great scientific revolution in the world.

Atlantis lay on the Atlantic ocean, inhabited by thousands of years of highly developed civilization. About 11,500 years ago, she suddenly sank after dramatic changes. There are many stories around Atlantis, its very location has been defined on the globe in several places. The most likely place on which she was located is the Atlantic Ocean. It must have been a very large continent and reached its size from the western shores of Europe to the Caribbean Islands.

And who were the Atlanteans / Atlantis?

The first Atlanteans arrived 50,000 years ago from Heaven. They resembled a human appearance. Cosmologists claim that originally they were inhabitants of the Liry/Lutnia system (constellation). They knew the Annunaks and are mentioned in a hidden way in the Bible - Genesis. They had a coded life for 800 years, they were very tall.

Archaeologists shocked human skeletons found on Earth, from 8 to 12 feet (2.66 - 4 m) high. The Spanish invaders on the American continent left notes that very tall, strange blondes with blue eyes were running in the Andes mountains and informing the inhabitants of the invaders. Scientists put forward the hypothesis that man has already manipulated genetics and created such a man for his own purposes as a worker.

Similar experiments were done by Annunaki.

The Bible mentions the supreme man, Goliath, about the height of six cubits and a span. According to the calculations of today's scientists, it was approximately 11 feet (3.7 m) high.

The highest man in the world living in the 20th century was Robert Wadlow from Illinois, he died in 1940, he was 2, 71 m high. Currently lives in the world, in Ukraine Leonid Stadnyk, is considered the highest in the world. It is 2, 54 m high.

At the beginning, the Atlanteans were less material, sensitive, and developed more spiritually. Later, they started to develop technology and reached its higher and higher level. They even managed to control the weather. Thanks to this, they harvested wonderful crops from their fields and gardens. They were also famous as good surveyors, they mastered volcanoes. Their admirable work was the creation of beautiful fire fountains from volcanoes. They had a lot of time, they were considered immortal.

The Atlanteans used crystals and developed their technology more and more deeply and finally lost their energy balance and led to their own catastrophe. Their knowledge of crystals with which they could focus a high energy beam like a laser served them for various purposes. The Pyramids of the Atlanteans are mentioned, although today's world does not have enough knowledge about them. But it is known that with them, the energy of crystals from the pyramid to the pyramid was transmitted. The pyramids were just stores of crystal energy. They used large amounts of pure quartz, made pyramids of various sizes, turned them up, consisted of four and six walls. They were also in different colors: white, pink, coral, gold, yellow, black, depending on what purpose they were intended, for example, pain relief, surgery in the body, deeper penetrating the human organs, the more they strengthened the vibrations of crystals. One of the most frequently used stones was ruby. He helped with emotional and spiritual problems, and black crystals were used for the same purposes. They served as strong protectors.

In order to restore their own vitality and youth, the Atlanteans used daily meditations in the vicinity of a circle made of 6, 11 or 24 stones of various types and in their hands they kept pure quartz. They could strengthen the crystal energy with the rays of the sun, moon and other stars.

A large fire crystal served like a central broadcasting station, using it to communicate in a similar way as we are currently communicating via telephone or radio. With the help of crystals they could leave their own body, they could never come back to it, they moved into higher dimensions. They were famous for dematerializing things and even terrains. The famous Philadelphia experiment conducted by Albert Einstein was modeled on the knowledge of the Atlanteans. They also decorated their own houses with crystals and served various purposes. As long as the energies of the crystals had the proper vibrations, everything went smoothly, but one day they sounded too high and the earth moved. It released volcanoes and these began to dissolve the mountains and caused the flooding of Atlantis.

According to Cayce, Atlantis had twelve communication points on Earth, very strong spirits, they were like temples and they served to communicate in other more distant groups. In the temples the oldest souls, highly advanced spiritually, served as priests. Masters trained properly for this role and had great power. They continued to develop their own consciousness and led less conscious people along this path:

through prayer, meditation and communication with the Higher Springs.

High priests were able to travel from Earth to higher dimensions, they knew the power of their own mind, they also had high knowledge of diseases and their sources of insurrection. They could heal, they were not foreign to the bodies in the human body. It is said that the Atlanteans have reached the peak of evolution. They developed their mystical abilities very highly and joined the cosmic source.

Why, then, did they fall?

They used cosmic knowledge but they were not completely overflowing with Divine Energy. They used their knowledge for their own satisfaction, showed their power and became richer by manipulating other weaker units. Sometimes they did not care at all what cost their behavior was to others.

The Great Spiritual Hierarchy reminded more than once of Atlantis priests and all those who imitated them, that they are not following the right path and that the priests are forgetting, instead of serving God only interfere with his pure intentions, they create devilish powers in the world. They admonished them many times and it was not foreign to priests that Atlantis could die.

The Atlanteans have achieved powerful technology. As they say, the Celestial Masters of fire crystals were stored in the pyramids of Atlantis. The crystals were reinforced with amber. Thanks to the pyramids, communication between various points on Earth developed. They did amazing things and strengthen their own mental powers. In the assumption of the Ascended Masters, it is harmonizing energy, developing mystical abilities, but it is to serve all of us so that man can broaden his consciousness and connect with the Supreme Source.

Energy harmonization and higher spiritual development have always been put in first place. This process is carried out by twenty-two High Masters on Earth who have high vibrations, so-called brilliant rays. Through these rays they can send energies to other people, strengthening their spiritual awareness.

The priests in Atlantis used high energies for their own purposes, they did not strengthen other people. Large crystals had the power to produce brilliant rays. Priests were losing their purity and their consciousness did not focus too much on spirituality, but they combined with the energies of the earth and eventually led to disaster. Before the Holocaust itself, humanity was not able to achieve too high a spiritual level. It is very important that everyone understands one big secret - the purity of their own mind, body and spirit. Only this great purity is able to elevate man above the earthly plan.

We have the exact picture painted by the hands of the Atlanteans, their power and the fall of this great civilization at the moment when they themselves have reached the peaks and cosmic powers but have lost their beautiful divine qualities. They turned away too much from the Divine Laws. The Atlames also produced synthetic crystals. Many people in the world say that synthetic crystals could have been the source of the destruction of Atlantis.

It is also important that a man working with crystals never think negatively. His thoughts quickly capture the crystal. Edmund Harold in his magnificent book "Treatment with Crystals" writes:

During Atlantean rule, priests who were pioneers in the development of power with crystals neutralized their mental, physical and spiritual fields. Crystal energy pulses with energy from the person who uses it.

Negative thoughts can cause a big problem, they can be a great destructive weapon. Atlanteans abusing their authority were called the sons of Belial. According to the Old Testament, they served Satan, they were rebels of cosmic law:

The wicked men went out from among you, and they deceived the inhabitants of their city, saying, let us go and minister to other gods whom you do not know. (Deuteronomy 13:14)

Do not go in a foreign yoke with unbelievers, because what does justice have to do with iniquity or what kind of community between light and darkness? Or what agreement between Christ and Belial, or what department has a believer with an unbeliever? What arrangement between the temple of God and the snowmen? For we are the temple of the living God, as God said: I will dwell in them, and they will be my people. (2 Cor. 6:14 -16).

According to the Hebrews, Baal - zebub - the king of darkness, depicted as an individual constantly changing to destroy their beauty.

I will continue to take away the attitude of those who want to have an attitude that in what they boast are as we are. For they are false apostles, treacherous workers who only take the form of the apostles of Christ. And no wonder then, and Satan takes the form of an angel of light. Nothing extraordinary if his servants take the form of the servants of justice, but their end is as their deeds. (2 Corinthians 11: 12-15).

The encyclopedia states that Belial - the prince of hell is the commander of 80 legions of demons, the leader of the sons of darkness. (Jude).

The Bible often mentions Belial:

"Levi," he told his children to choose God's Law or work for Belial

Joseph from Egypt said, "They will be in the light of God when Belial will be in the dark."

"When the Messiah comes, the angels will punish the spirits of Belial."

Two brothers Moses and Aaron, known from the Bible, two opposites.

The satanic Bible is called Beliala - the four-crowned prince of hell. Each crown represents a different earthly element. Belial is the master of life on Earth and he loves pleasure, sex, he takes care of his own "I", he likes to control others and he develops his own "superego".

Belial uses different tricks to push people off their right path.

His greatest achievement is to let him just not believe.

The second is that the Mercy of God is so great that a man will be forgiven of all faults and even for the greatest life deeds God will give him the same place as the one who was always just and right.

Yes, God is Merciful and forgive all guilt but only to a man who fully understands his mistakes and corrects the wrongs done to others. His soul must be clean.

The best example are the stories of some saints:

"They did wrong

But they understood their own mistakes

They turned back from fake roads

We have the time of Mercy and it is a great time to repair ourselves, and the time of justice comes in a similar way.

In Atlantis, the Celestial Masters warned the priests, admonished them and they announced a great fall to this land if they did not turn back from the false roads in time but their voices were ignored.

The day of justice has come

Many Ascended Masters now provide information about this missing continent. Edgar Cayce was one of the many channels through which we got the knowledge of the long lost Atlantis.

Source - http://www.vismaya-
maitreya.pl/zakryte_zagadki_atlantyda.html

* * *

In addition to Atlantis, the esoterics are still talking about another lost continent - Lemuria, which is identified by some with the lost continent Moo (Mu) in the Pacific and by the other with Lanka in the Indian Ocean - their remains would be the islands of Polynesia and Sri Lanka and Seychelles, Maldives, Comoros, Adamanda and Nicobar. Here is what esoterics write about this topic:

Lemuria (Mu):

LEMURIA was an ancient civilization that existed before and during the existence of an ancient continent called Atlantis. It is believed that Lemuria existed to a large extent in the South Pacific, between North America, Asia and Australia. To clarify the information, it is worth noting that Lemuria is sometimes referred to as Mu or Homeland (from MU). Initially, the concept and name of this land were proposed by August Le Plongeon in the nineteenth century - a traveler and writer who claimed that ancient civilizations, such as Egypt and Mesoamerica, were created by refugees from MU - a continent that is located in the Atlantic Ocean. Then this thread was popularized and developed by James Church- tor (1851-1936) in a series of books, he claimed that he was once in the Pacific.

The first book of Churchward, "The Lost Continent", was published in 1926. Churchward said that a certain Hindu priest showed him ancient plaques created by the Naacals - an advanced civilization inhabiting the continent of Mu. This land, according to his messages, was to have a length of 9600 km and extend from the point lying north of Hawaii to Easter Island.

According to Churchward, volcanic eruptions, earthquakes and powerful waves destroyed about 12.000 years ago. The remains of the continent are to be islands scattered in the Pacific Ocean.

There is also another hypothesis about Mu. This continent was supposed to have emerged in the course of Earth Perversion 52 thousand years ago, and the sinking of the continent was the result of another reversal, about 24 thousand years BC

The population of the continent according to this thesis moved the continent re-emerged as a result of changes in the lithosphere continent. A new land appeared on the surface as a consequence of changes in the magnetic field in the planet's core. The continent was Atlantis. According to the authors who created this thesis, the Atlanteans were to be descendants of the inhabitants of Mu.

Another person who was interested in the subject of the oldest culture on earth, such as Lemuria is Frank Joseph, an American explorer of the past, editor-in-chief of "Ancient American" magazine, a longtime researcher of the Burrows cave collection, author of the book "The Holocaust of Atlantis" and "Atlantis Survivors" ". Joseph published his discoveries related to Lemuria in the book entitled "Mystery of the Oldest Culture on Earth", where he presented m.in. underwater testimonies of the existence and destruction of Lemuria civilization. Based on the latest discoveries of underwater archeology, peculiar hieroglyphs and symbols as well as ancient records, he recreated the image of a lost world that, like Atlantis, sank as a result of a terrible catastrophe. Also Frank Joseph claims that the surviving Lemurians reached other continents, imparting their knowledge to the peoples of Asia, Polynesia and the Americas.

The existence of Mu's continent has, however, been questioned. Scientists today reject the concept of MU (and other continents, such as lost Lemuria) as physically impossible, because they claim that no continent could be destroyed by a cataclysm, especially in such a short period of time. In addition, as the researchers say, there is a mass of opposing archaeological, linguistic and genetic evidence that the ancient civilizations of the New and Old World have a common origin. Due to these "facts", this theory has been explained by scientists as untrue. Therefore, it is considered today as a fictional place, and books on this topic are generally found in the New Age movements and so-called groups dealing with "Religion and Spirituality".

But can we trust these scientists to the end? I would like to remind you that some time ago the scientists did not take into account the existence of Troy. It was thought that this is only the imaginary history of Plato[21], written to strengthen the bravery of the Greek army. However, years later it turned out that Troy existed and there is clear evidence for it. Why am I writing about this here? Well, because it was Plato who mentioned Atlantis, and he touched on this subject in antiquity.

However, going further along the trail of information about Lemuria, one should mention the visionary of world-renowned Edgar Cayce. In one of the records, during his vision of the place of sinking the continent of Atlantis, Cayce mentioned that the civilization existing there was the third civilization in the world. And just before the Atlantean civilization, Lemuria existed.

Currently, Lemuri can only learn from the so-called "messages" of beings who claim to be people living in those times. One of such beings is Ramtha, who claims he lived 35 thousand years ago on the Atlantic continent as Lemurian.

In the peak period of civilization development, Lemurians evolved a lot and were at a very high level of spiritual development. However,

---

[21] Homer told the story about Troy in the poems "Iliad" and "Odyssey", not Plato.

the Atlanteans, after developing the technique and moving towards intellectual development, hated and despised Lemurians. To the extent that Lemurians were treated worse than dogs. As Ramtha describes, "the dogs on the street lived better than Lemurians because they were at least fed".

About Lemuria from another source

The development and collapse of Lemuria is nowhere documented. There are no records or other records that would say that this land really existed. Scientists have already assumed for a long time that there must have been another land long before Atlantis. Lemuria also known as the Pacific, MU and what he called her Cayce Zu or Oz was not similar to Atlantis. It is assumed that there are still places on Earth, the remains of this great continent. Many islands in the Pacific, for example, the Great Island of Hawaii and the Easter Islands are attributed to the Earth of MU. Many scientists believe this. On the Easter Islands is wondering where the colossal stone figures came from. Maoris in New Zealand still remember that a long time ago there was a sunken island called Hawaiki. There are many dates in time that are attributed to the existence of Lemuria, from 75,000 years to 20,000 before Atlantis. There are also assumptions that the Lemurians and Atlanteans were in contact with each other during a certain period of their common existence.

Bławacka wrote about Lemuria.

She accurately described many secret symbols from the land of MU. MU August Le Plongeon (1826 - 1908), a scientist studying the Mayan culture in the Yucatan, remembered the Earth. After translating the Maya texts found in the old ruins on the Yucatán peninsula, it was obvious that there was another land before the culture of Atlantis and ancient Egypt, but the researcher considered them fairy tales, they spoke of an old continent called MU.

HP Bławacka published the book "Book of Dzyan" ("The Stanzas of Dzyan") in 1880, a set of cosmologies, describing seven races of man. In his book, he presents to the world for the first time Lemuria and its inhabitants - the third race of people on Earth. He describes them as very high-rise beings, hermaphrodites (having both male and female reproductive systems, produce eggs and sperm in one body, as well as many plants and some of the lower group animals). Mentally, they were not very developed, but more purely spiritual.

HP Bławacka also gave the reason for the Lemuria wreck. A certain group of Lemurians began sexual practices with animals. Lemuria was sunk, and in this place a new fourth race of people began to appear, they were the Atlanteans.

In 1894, Frederic Spencer Olivier in the book "A Dweller Two Planet" writes about the sinking of the continent Lemuri and remembers Mount Shasta in California as being a remnant of this old land. Similar information is provided by other groups, including:

Church Universal and Triumphant

The Temple of the Presence

The Hearts Center

In addition to Lemuria, the name Kumari Kandam is also mentioned - the Kingdom in southern India, an existing civilization about 50,000 years ago. It is difficult to determine whether it was a civilization with Lemuri or lived simultaneously at the same time.

So much science - what do the mystics say?

Mystical sources report that the Lemurian race was a mixture of mainly Sirius, Alpha Centaur and less of other planets. All these races mixed together on Earth during the formation of Mu's civilization. Lemuria is portrayed as a heavenly and magical land, for which for a long time people did not know major concerns. 25,000 years ago Lemuria and Atlantis were the two largest Earth civilizations. They fought for each other's own views. Both civilizations had very different ideals. Lemurians believed that other less-developed cultures should go to loneliness to continue their own evolution, in their own silence, so that they could better understand their own way of life. The Atlanteans believed that less developed cultures should be controlled by two more developed civilizations. This difference of views caused a series of thermonuclear wars between Atlantis and Lemuria.

When the wars had expired and the war chicken had died down, there was no winner. During these wars people, highly advanced, went back to a completely low level of development. Both Atlantis and Lemuria were victims of their aggression and greatly weakened both continents.

During Lemuria, California was part of their land. In the area of Mount Shasta they built up their culture and built a powerful city - Telos. It stretched on very large terrain, not only around Mount Shasta, but along the whole of today's west coast of North America, it stretched all the way to British Columbia. Telos means communication with the soul, unity with the soul and understanding of the soul.

Lemuria is considered the main land of man. She died very quickly in one night. Everyone slept and did not even notice what had happened. Lord Himalaja, the Ascended Master, in 1959 by Geraldine Innocenti, he provided information about Lemuria and its disappearance. Before she sank, the priests and priestesses from the Temple warned against the impending cataclysm. Many Saints and Holy Fire managed to move to Telos. They also managed to evacuate a lot of people to safe places. Holy Fire was also sent to the continent of Atlantis, placed in special places and secured for a long time to be used for spiritual purposes.

Before the sinking of Lemuria, the priests and priestesses returned to those places that were to disappear to help people in this difficult moment, so that there would be no fears and fears in them. They did a lot of their work with energy, they wrapped the bodies of people with their own aura to save their emotional bodies.

Lord Himalaja in his transmission entitled "Liberty Bridge" from 1959 says:

Many members of the clergy have formed small strategic groups in different regions. They prayed and sang to heal emotional wounds in the etheric bodies and save the memory that was encoded in the cells. During this action and dedication, the priests chose to stay with the groups and until the very end. They supported people full of fears. They were praying until the waves of water filled their mouths. During their prayers, people fell asleep, the water covered them suddenly. In the morning it was blue sky, it all ended. Lemuria rested on the bottom of the Pacific. None of the priests escaped, nor did any of them fear. Lemuria has gone dignified.

At the same time as Lemuria was leaving, she also began to sink in Atlantis. It quickly lost significant parts of its land. This process lasted about 200 years until Atlantis finally sank completely.

Both these continents were already under water. The remaining small islands and the people who lived on them still remembered these events as well as nuclear wars. It took a long time before they healed their emotional wounds. After the sinking of both these continents, the atmosphere drastically changed on Earth. There was great cold, there was no sun in the sky and the atmosphere was still contaminated. There were very few plants growing on the Earth and there was not enough food. A large number of animals have disappeared.

For: http://www.lemuria.pl/LEMURIA.html

* * *

Of course, apart from the whole biblical-demonological gibberish that is so in the history of human civilizations as the fist to the nose. After all, the Bible itself has been re-edited and censored for purely ideological reasons over almost 2000 years. They removed everything that did not suit the fathers of the Church, and therefore referring to it as a source is burdened with a very high risk - that is, in all this, some rational grain of truth. We quote this only as a curiosity, but on the other hand, it is in the Bible that there is much information distorted and distorted by numerous translations and interpretations that may pertain to past events related to the extinction of the previous civilization of the Atlantean Empire but already mentioned castration over the centuries prevents the exact location of these events in time. Information contained in it can be treated only as so-called signaling information - elemental information that needs to be verified and deepened.

Much of this text is about energy and the unknown energetic properties of crystals that allegedly intended to produce energy for the needs of the Atlantean civilization. We believe that it was electricity, which was drawn from crystals, eg quartz, using the piezoelectric effect. There is nothing mysterious or unknown in it. Atlantis was a volcanic island, like today's Iceland, so the geothermal and hydrothermal energies of the Atlanteans had holes in their noses, and the crystals could serve them more as electric energy devices than they produce.

As we have already written many times, Atlantis stood on the so-called hot spot - the hot spot under which the plume was locatedor if you prefer a fountain of hot magma from inside the Earth. When he disappeared - the island, which constituted the seabed washed up by Iceland, it sank to the bottom of the Atlantic. Its traces can be seen at the bottom of the ocean just vis-à-vis Gibraltar, where there is a particularly mountainous area corresponding to the description of the Platonic island. Recently, regular lines were discovered on the seabed suggesting the existence of some artificial structures - perhaps irrigation channels or something like that. And what's interesting is that scientists have taken water on the mouth and do not comment on this discovery.

Now, let's go back to the megaliths. Robert Leśniakiewicz claims explicitly that in Poland they could not have been created earlier than 11,000 years ago. Why? And because it was the Ice Age and a large part of our country was covered with ice cream. Only south of Poland was free from them, but if any circles or other stone constructions were built there, they would break down under the impact of the earthquake haunting the earthquake, which often reached 9-10 ° R even in historical times! Drawing up a book about the Moon Cave together with Dr. Miloš Jesenský, he drew attention to this phenomenon, which occurred and still occurs in Central Europe. The presence of earthquakes also explains why northern Poland was built in such a small circle of menhirs. Well, it was done just because they were not destroyed by strong tectonic shocks.

Returning to the glaciation, in the Pleistocene the area of Poland was repeatedly occupied by a continental glacier, which periodically melted, so most of Poland covers Pleistocene glacier and glaciation sediments.

The glaciation of Narwia / Günz - about 1 million years (MA) - 0.6 MA ago is considered the oldest glaciation; the forehead reached the northern foreland of Wyższa Lubelska, the region of the mouth of Pilica and the Płock region, from where it bent towards the north towards the Lower Vistula Valley; the glacial clay occurring in the Gałachów region near Modlin is a trace of this glaciation. During the Interglacial of the Podlasie, the rivers modeled deep valleys filled with alluvial sediments, mainly sandy-gravelly, sometimes mules with flora debris.

During the Nida / Mindel 0.5 - 0.4 MA glaciation, the ice sheet leaned against the northern slopes of the G. Świętokrzyskie, covering the Wzniesienia Łódzkie and Wyż. Lubelska ranges; this glaciation is traced to the oldest level of boulder clay, preserved particularly well in the western and northern part of the Świętokrzyskie Mountains, near the Kleszczów canyon near Bełchatów and in the deep valleys on Wyżna Lubelska; the accumulation of the oldest loess in Poland, as well as mules and loamy clays is also connected with the Nida glaciation period. The sediments of the Lesser Poland (Przasnysz) interglacial are preserved only in a few places. These are river, lake and swamp settlements with the remains of fauna and flora, occurring on Wzniesienia Łódzkich, Niz. Mazowiecka and Wyż. Lubelska.

During the glaciation of Sanu I / Riss 0.2 - 0.1 MA, the ice sheet crossed the belt of the central Polish highlands reaching the foreland of the Sudetes and the Lower San Valley. The trace of this glaciation is the level of boulder clay by approx. from 5 to 30 m. During the Ferdinand interglacial, the ice sheet completely resigned from the area of Poland, erosion took place; sands, gravels and residual pavements were formed, and in the lakes also mules, diatomaceous earth, lake chalk, gyttia and peat.

During the glaciation of San II, the glacier's range ran along the northern slopes of Roztocze and the Carpathians, from where its boundary ran along the Sudetes; the deposits formed at that time include tills, as well as sands and gravels, with a total thickness of up to several meters; valleys of rivers flowing down from the Carpathians were then buried up to a height of 90 m above the level of their contemporary troughs.

In the interglacial of the great, which is clearly bipartite, river sediments were formed, as well as lake sediments - mules, clays, lake chalk, diatomaceous earth, peat; these settlements are known from many places, both in the Polish Lowlands and on the highlands of the Carpathian and Sudety Mountains.

During the glaciation of the Liwiec, the continental glacier covered its north-eastern Poland, while its forehead reached the basin of the Liwiec River to the parallel of Warsaw; sediments of this glaciation are represented by boulder clays, mules and clarets formed in the shanty lakes formed before the forehead of the glacier; during this glaciation, the oldest loess in southern Poland was created.

In the Zbójna interglacial, erosion processes dominated (erosion of older glaciation deposits); in many places alluvial and limnic deposits of this interglacial have been preserved.

During the Odra glaciation, the ice sheet leaned against the Sudetenland, slightly entered the Moravian Gate, and then its forehead ran north-east, bypassing the East Świętokrzyskie and leaning against the northern edge of Roztocze; the maximum range of this glaciation is indicated by the frontal moraine stripes, as well as boulder clays, sand fields, kame hills and sows on Wzniesienia Łódzkie, Mazowiecka Lowland, Podlasie, as well as at Śląska, Wyższa Małopolska and the foothills of the Sudetes. In Małopolska and Wyżna Lubelska, sediments of loess occurred. Organogenic, river and limestone deposits formed in the Lubawskie interglacial; in the loess of southern Poland there are fossil soils from this interglacial.

In the Warta glaciation, 3 stages are distinguished; during the maximum stadium - the Pilica stadium, the ice sheet reached the Lower Pilica Valley, crossing it slightly east of Warka; east of the Vistula, its range is determined by the water section between the Wieprz, Krzna and Bug basin. To the west of the Vistula, the glacier's forehead formed a huge lob in the region of Łódź, then ran westwards; each stadium was left behind, apart from boulder clays separated from each other by sandy-gravelly water-glacial sediments or silt and courtyard clays, frontal moraine belts and sand fields.

In the Eemian interglacial, the waters of the sea floated into the region of the lower Vistula, while river and lake settlements formed in the remaining area of northern and central Poland.

Glaciation of Wisła/Würm 0.07 - 0.01 MA, divided into three stages: the stadiums of Toruń, Świecie and the main stadium. During the Torun stadium, the ice sheet twice entered the area of the Lower Vistula Valley, previously occupied by the sea bay. During the Świecie stadium, the ice sheet once again entered the Lower Vistula Valley, and perhaps also the area of Warmia and the northern part of the Masurian Lake District. In time the main stadium there are two phases: Leszczyńska and Pomorska; the glacier reached the maximum range during the Leszno phase, its forehead ran from Zielona Góra through Leszno, Września, Konin, Płock, Niedzica and further east. Before the forehead of the glacier, Pradolina Głogowska was created, which along with the Pradolina Bzura-Ner and the Warsaw-Berlin Pradolina flowed westward from the melting ice sheet. In the Pomeranian phase, the continental glacier covered only northern Poland, entering south of the valleys of the lower Vistula and Odra. The water that gathered before the forehead led westward the Toruń-Eberswald Pradolina.

During the glaciation of the Vistula and the glacier recession from the area of Poland, the leading moraine hills, numerous ozes, kames, sand fields, as well as gutter, dam and lake lakes were created. Lime sedimentation occurred in southern Poland.

During the Pleistocene glaciations, mountain glaciers in the Tatras and, to a lesser extent, in the Giant Mountains were created. The remains of them are numerous circus lakes, glaciers, side moraine embankments, ice scrapes and glaciers.

(Source: http://encyklopedia.pwn.pl/haslo.php?id=4575162).

So what's the result? And the fact that megaliths in our country could not be built in Atlantis, because the territory of Poland was covered with snow and ice-melting ice, and up to the Beskids. They had to be created later, when these areas were possible to be inhabited by people. Megaliths in Poland are arranged very interestingly. Megalithic monuments in Poland are associated with the occurrence of two cultures: the funnel-shaped culture and the later culture of spherical amphorae.

The first of these is the funnel-shaped cup culture. This culture in the third millennium BC was spread over vast areas of central and northern Europe. Its remnants include various megalithic tombs. From the area of Poland known for this period, long, bezkomorowe kopce ziemne, called kujawskie mounds. They had the shape of an elongated trapezoid close to a triangle, the width of the base from 6-15 meters, length from 40 to 150 meters, and the original height is estimated at approx. 3 m. The bulwark was made of large boulders. In the widest part of the building, called the base or the front of the tomb, there was a single grave, sometimes also more burials. Between the grave and the stones of the base there was sometimes a sanctuary (tomb temple). The Kujawy tombs were erected in the period from the fourth-third millennium BC. The most famous ones are currently located in Sarnowo and Wietrzychowice in Kujawy, as well as in the vicinity of Łupawa in Pomerania.

The second is the culture of spherical amphorae. In the second half of the third millennium BC, completely different megalithic tombs, characteristic of the spherical amphorae culture, began to appear in Poland. These were box tombs made of large slabs and covered with a large boulder or boulders. Their length ranged from 2.5-6 meters, and the width was about 1-2 meters. These boxes were buried under the surface of the earth. They found collective burials, sometimes also animal sacrifices.

Megaksylons. The megaksylon tomb referred to the megalithic tombs of the Kujawy type, but the pillar of its construction was large trunks. In Poland, such tombs were found in Słonowice, in the south of the country. It is not difficult to explain why the builders of megaxylons have gone from using stone to wood - in Kujawy and Pomerania there is no shortage of postglacial erratic boulders. In the south they are rare. Attachment to the idea of a megalithic building forced the use of another material. The walls of the Słonowice megaksylons were made of a huge palisade of buried trunks with a diameter of about 30 cm each. The width of the grave was about 5 m, and the depth (above ground level) was about 3 m. Megaksylon was leaning towards the west. Of course, only the fundametra grooves have survived to this day, in which the trunks of the tomb have been set, with well visible, much darker wheels - a remnant of the spread vertical beams. After spreading the wood, the embankments were washed away by rain.

However, burials have been preserved. To date, six megaksylons have been discovered in Słonowice, located in parallel, with common features, but not identical and differing in size. Only two of them are accompanied by ditches, from which land was chosen to fill graves. Perhaps, therefore, the others are empty. The burials do not contain any interesting equipment, the bodies have been covered with stones and earth. Sometimes, researchers find a vessel or copper products. Research in Słonowice began in 1979, is currently underway, and their end is not visible yet. More on this subject can be read on the website Słonowice on the river Małoszówka, dedicated to this excavation. (Wikipedia) And so all within the range of glaciers that hit Poland. Incidentally, the volume of only one tomb in Słonowice is equal to 1.5 volumes of the Great Pyramid!

As for the stone circles, Cromlechów, they only occur in Pomerania and again, they are not in the southern parts of Poland free of glaciers. What does it mean? Well, it's just that they were built after the last glaciation - after the catastrophe, which plunged Atlantis into the depths of the Atlantic Ocean. This is an absolute certainty. Few realize what we are dealing with here. Even if some civilization erected its most massive and huge buildings like pyramids during interglacials, each subsequent glaciation would bring annihilation. Hardly anyone realizes what these glaciers really were. Something similar today is only in Antarctica. It was a slow but inexorably steady ice cube in Europe, 2,000 m thick, capable of crushing, compacting and disperse their ruins over hundreds of square kilometers, and the most powerful even stone blocks to coat and polish, that they would resemble ordinary erratic boulders, which are abundant in the Polish Lowlands. And that may be (but they do not have to) traces of THESE buildings of ancient civilizations. All others have been created in the last interglacial, in which we are also.

What is the conclusion? Well, that all stone buildings of Pomerania cannot be older than 11-12 thousand years. And what's more, all of them were created after the cataclysm, which absorbed and plunged in the sea waves of the Atlantean Empire. The collapse of Atlantis triggered the entry of the Gulf Stream into the northern part of the Atlantic and, as a result, the huge amount of heat that melted the glacier. Let us not forget that the sea currents are the "transmission belt" of heat. If 35 MA MAY have no sea currents forced to circumnavigate Antarctica without supplying heat from the equatorial belt, then Antarctica would not have an ice shell up to 5,000 meters thick today, and it would grow lush forests like in Canada or Siberia because it would be comparable climate and in the worst case. Megalithic buildings could only be built in the south of Poland, and this is what he and Dr Miloš Jesenský based their hopes on finding the Moon Cave - if it exists on the Polish-Slovak border. Then, of course, there were mountain glaciers, but that was not what the Polish Lowlands did. Here you could live and build buildings. It was also possible to build defensive installations, as some want, or e.g ballistic rocket silos - how they want the greatest fantasy. The Moon cave, according to the words of Dr. Horák, was to be at least 20,000 years old, and therefore it was created during the last glaciation.

Vislan/Würm, when significant parts of Poland were covered with ice sheet of the last glacial. And what's more, it could not be created (assuming it was artificial and intended) in the High Tatras or Low Tatras, because these areas were covered with mountain glaciers. Therefore, this is another argument that it was created - it was built - in the Beskid Sadecki on the Polish side or Lubovnianská Vrchowina or Levočský vrchy on the Slovak side. But let's return to megaliths. In one of my articles, Robert Leśniakiewicz wrote that they constitute a distant memory of what once could have existed in Atlantis and its colonies on all continents. They are associated primarily with some astronomical data. And the fact - in most cases this is how it is. Of course, the "scholars" "explained" that this is related to the solar cult and cult of fertility of the Mother Earth and was used to calculate the equinoxes and summer and winter solstices, itede itepe. But does it explain everything?

Consider for a moment: what does every technological civilization need?

Of course, two things: raw materials and energy. It was no different with Atlantis - she needed both. Without it, it simply could not exist, and as Aristocles aka Plato writes - to be hegemony over the great tracts of the world. It has always been so, it will be so. Raw materials and energy - especially energy - is an exponent of the power of a given civilization.

And another question: what was the energy? Of course, the cheapest and the most accessible - the energy of the Earth's interior. Like in Iceland. But that was the Achilles heel of this civilization, because it was only in Atlantis. And what about the colonies in areas where volcanism was minimal or none? The Atlanteans had to deal with each other and dealt with finding a second, powerful source of energy. No, not atomic - this was marginalized as dangerous to the environment. The Atlanteans knew her, but they did not use her. Therefore, the Moon Shaft could be such a trace of the former mining works that were abandoned. So the latter was solar energy!

And here we come to the heart of the matter. Why did our distant ancestors put stones in circles instead of squares or rectangles? Well, because they remembered that their Atlantean Masters and Teachers were arranging the mirrors and collectors of their solar power plants in circles.

Let's look at the designs of such power plants - they consist of a circular system of mirrors and a central collector, which brings solar light to converters, which in turn convert thermal energy into electricity. Something like that showed Breck Eisner in the film "Sahara" (2005) according to the novel by Clive Cussler. Now it is obvious that such power plants had to be precisely oriented towards the world sides and their location did not have anything accidental - they had to be oriented so that every caloricity of solar heat could be used optimally. I think it is clear and obvious.

Technical knowledge about solar power plants in the degenerate form of religious beliefs survived until the time when stone circles were built, orientated according to the world - this time not to get electricity from them, but to set some astronomical data - hence Stonehenge phenomenon whether our circles in Odry or Grzybnica and the exact convergence of their location with the location of solstice points and other important for the then primitive hunting and agricultural clusters that remained after the destruction of the powerful mother civilization. In short, we are dealing with a memory from the distant Past of our species. A memory that counts at least eleven thousand years. And it is true that megaliths have something to do with energies - they are ineffective models of something that existed in the distant past. All this was later distorted and distorted by oneiric-delirious delights of various religious bewilders - hence all metaphysical nonsense, which have no power in reality. Lack of knowledge caused fear of the world, and this in turn gave birth to magic and religion. In this and other cases, one should always approach the issue according to the motto of Aristotle, which is expressed by the sentence - Amicus Plato, sed magis amica merita est , and what does not always go to scholars and scholars who always want in the exact sciences to find some place for some unrealistic of being and its causative abilities.

Did the Atlanteans know nuclear and thermonuclear energy? I think they did, but they did not use it because of its side effects that they had to know. Everywhere in the world there are traces of mining uranium and uranium-thorium ores, which are several or even several thousand years old. Dr Jesenský wrote about this in his great work "Gods of Atomic Wars" (Ústi nad Labem 1998), in which he claims that uranium compounds - and enriched ones! They were already interested in deep Antiquity. Is this another proof for the past in which uranium compounds were used for industrial and military purposes?

I think the hypothesis set here is verifiable. Above all, you need to search for artifacts in areas where Atlantis colonization was possible on both sides of the Atlantic Ocean - especially in the areas of Amazonas or Brazilian selva, where there are no volcanoes and Sahara sands (which was then a rich and fertile land). Personally, however, I bet on Europe and Africa - because it was there that there was a fashion for putting cyclopean boulders in characteristic circles, which were then worshiped. It is obvious - solar power plants were something extremely important for Atlantis and that is why they were so carefully looked after. They were a sine qua non condition for the existence of their civilization. When they ran out - then it ended.

And why did you run out? Well, it was missing because there was a catastrophe that devoured Atlantis and its inhabitants. As has been said before - Atlantis was on a fragment of the seabed, drawn by a plume of hot magma from inside the Earth - like today's Iceland, Hawaii or the Azores. This "hot spot" was a perfect place for people to settle: fertile soil, excellent climatic conditions - similar to those that prevail on the Costa de la Luz in the Spanish Andalusia, the abundance of hot springs and finally the warm current flowing the island from the south - all this a real paradise on Earth. According to Greek myths, it was here where the Garden of the Hesperides was located, where the powerful Heracles and the cunning Ulysses-Odysseus arrived. And it also agrees with Plato's description.

What scenarios of this doom could have been possible? First of all, the violation of the continuity of the earth's crust, which caused a volcanic explosion comparable to the explosion of the Lake Toba Superwolk - which took place 70,000 years ago. Please compare this to the reader with the period of overlap of Europe with ice IV Vislan / Würm glaciation. Somehow it fits one to another, does not it? However, this does not match the picture of the situation in Europe, which has become free of ice cream.

The second concept is by Otto Muck - a hit in Atlantis of an asteroid and a repetition of the pre-winter winter from 64.8 million years ago, when dinosaurs died out. In both of the cases mentioned here, volcanic eruptions could have occurred and, as a result, Atlantis flooded the waters of the ocean. The hot plume no longer pushes the magma that has passed through volcanic craters or the impact crater, and the fogging of the atmosphere with post-treatment or post-thermal dust lasting several years has resulted in lower efficiency of solar power plants, which have ceased to be profitable. And when the dust and fumes disappeared after the Great Catastrophe - there was no one to repair, set or use. There are only memories and technical descriptions that, as time passed and people began to be primitive, first became a collection of legends and applications, and then a system of religious myths. Their fading echoes reached the Greeks, Egyptians, Chaldeans, Phoenicians, Jews and other peoples who created their own beliefs.

Of course, everything I wrote here is only speculation. However, I think that these hypotheses are much more likely and sensible than stories about gods or aliens who came from heaven and behaved on Earth - not-divinely. I would even say that they were unpredictably human, envious, jealous and inconsistent. But they also loved and lived humanly and though immortal - they were associated with people and had offspring with them. It is obvious that they were only people who achieved a high degree of civilization, so that they could be considered gods for the primitive inhabitants of the Mediterranean, Black and Baltic basins. Just like we are for Aborigines or inhabitants of the Amazonas jungle. And perhaps our civilization will end in the same way as the civilization of Atlantis once, and then they will start to create their own myths about us and our lives.

And another circle of history will close, because let's not forget that history repeats itself as many times as it can. Or maybe rather as many times as we allow it.

CHAPTER VIII

Lunar Shaft - Ignis Fatuus?

Already after the publication of our book about the Moon Cave we received a lot of additional data about her and other strange phenomena related to the existence of various strange artefacts from the Far Past on the Earth, which could be related to the so-called "Platonic" civilizations of Atlantis, Mu, Lanka, Lemuria or even the Atlantic.

The search for the Moon Cave reminds us of the pursuit of the title erroneous ogre - when it seems that we already have an exceptionally "warm trail", it turns out to be more closely followed by a path leading nowhere.

Recently, we received a report from Germany by Ing. Dipl. Marco Hiltscher, who decided to find this cave and for this purpose he took three expeditions to the Beskids, where he expected to discover it for the world.

The original location

Let's try to analyze the material that we obtained from inż. Hiltschera. The first thing that raises the biggest doubts are the geographical coordinates of the Moon Cave. According to the accounts of Dr. Antonin Horak, who was the first scholar in it, the cave is located near the point described by geographical coordinates: 49.2 N - 20.7 E. The main question is: Where is this point?

Well, just - it is not very well known, because we do not know what we are dealing with. Is it N 49 ° 02 '- E 020 ° 07'? If so, where is this point? Looking at the map, we can see that this point is located between the village of Vikartovce and Liptovské Teplice in the Low Tatras, south-west of the city of Poprad. It would match the geological structure of the Moon Cave - a massif of carbonate rocks on a sandstone substrate - as it is, for example, in the Belianske Tatras. However, there are no two towns mentioned in the document: Plavnice and Lubocna and the village called Yzdar or Ždiar.

These conditions are met (again!) partly for two locations, the first of which coincides with the area of the Velka Fatra mountain range and is located in the Kral'ovan area. In fact - there are limestone rocks, a cave near the top of Šip Mountain (Dziura v Šipie) and two towns named Parnica and Lubochnia. It should be added that Šip mountain is located exactly between these towns. Only the third element is missing - a locality called Yzdar. In this area there are several geographical points with the core "Ždžar" in their names, but unfortunately, it does not match the given geographic coordinates.

So it should be read differently? But how? Maybe Dr. Horak gave them in the form we now use in GPS devices - and therefore not N 49 ° 02 '- E 020 ° 07', and 49 °, 2 - 20 °, 7 - and hence N 49 ° 12 '- E 020 ° 42'? This point would be located in the Levočské Vrh area - south of the village of Stara Lubovnia - situated at N 49° 18'34"- E 020 ° 40'44" and the village of Plavnica - N 49 ° 16'36 "- E 020 ° 46 ' 36 ", somewhere around the town of Jakubanske Kupele located between Nowa Lubovnia and Plavnica! All this sounds awesome, but again, there is NO place called Yzdar/Ždziar nearby because we assume that this is the name of this mysterious place. So the riddle remains a mystery.

The retreat of the insurgent battalion

The issue of the retreat of the detachment of the captain. Dr. Horaka after the skirmish with Germany is equally interesting. Perhaps the skirmish with the Germans took place in the areas given by the geographical coordinates, and then the detachment bounced back towards the north towards the border with the General Governorate (meaning Poland). Why just there, and not east - facing the Soviet army? The answer seems to be simple - he did not love Russians and did not love communism. But not only. At that time, this section of the border between GG and Slovakia was poorly manned by German troops and poorly protected at all. The reason was obvious - the Polish Resistance Movement wanted to have an easy stretch of border to overcome by couriers from occupied Poland via Slovakia to Hungary and further to England, where the headquarters of the Supreme Commander and the government of emigres were located. This section of the border was exceptionally active in this respect. It was followed by smuggling from/to GG and couriers of the Home Army, thanks to which to convey, if you prefer to pass through the cordon many valuable intelligence documents and materials on, among others German retaliation weapons. Following in this direction Cpt. Dr. Horak simply did not want to expose his unit to further losses, because he knew that there was little probability of encountering strong Wehrmacht or native fascists from Hlinkavej Garda.

Inż. Hiltscher conducted his research on the Polish-Slovak border. He does not give the name of the place where he stopped, but from what he writes about it, he lived in the vicinity of Piwniczna, from where he had an easy access to the mountain where the incrusted cave was located - he found it in the self-purchased one tourist guide. So far, the only mountain that fulfills this condition is Wierch Zubrzy - 861 m above sea level, in which a cave is to be found in the cave party from the east.

Beskid caves

We know from experience that the Beskid caves do not compare with their Tatra counterparts. These are rather holes / fissures called cave outcrops. Their length ranges from 2 to 25 meters. In the described case, it would be a cave with a length of up to several kilometers! Where does this application come from? And from here, that nearby - at a distance of about 3 km to the north-east there is the second interesting mountain - Pusta Wielka (Pusta) - 1061 m high. Personally, I am of the opinion that the search should take place there, despite the fact that Ing. Hiltscher believes that the holes he finds are leading to the cave system under the western part of the Sądecki Beskids. So far - as in the case of the Slovak Kopy near Kral'ovan - no trace of this system was found, and yet the probability of existence of such a system in Kopa is much larger than under Pusta. Kopa is a mass of carbonate rocks, while Pusta is made of sandstones.

Engineer Hiltscher gives one more interesting detail, because in the vicinity of both holes discovered by him, he found the presence of radioactivity, which would suggest that uranium ore could be found inside the mountain. Indeed, uranium ores in the Beskids were sought after by the Germans and Russians in the years 1940 - 1956, and, to the extent that the local people believed, traces of uranium and thorium ores were found in this area, including in Żarnówka (Beskid Makowski) or in the region of Ptaszkowa. But - as has been said here, these deposits are not worth exploitation. On the other hand, one detail agrees, namely - in this area there are sources of healing water - sorrel, hence the names Szczawnik, etc., which are saturated with carbon dioxide. Perhaps there are some underground waters saturated with, for example, hydrogen sulfide, which is as poisonous as CO_2. This fits the relationship of Dr. Horak.

...and the Lemkos

Even during the Second World War, the Lemkos' trips to Soviet Ukraine and the interior of the USSR began, initially voluntarily. With time, resettlements became more and more intense (under the influence of Ukrainian nationalist propaganda) and forced (an interstate agreement between Poland and Soviet Ukraine on the exchange of people). However, the "Vistula" Action, due to the accusation of the Lemko-Ruthenian population to foster the rebellion caused by the UPA in 1947, caused the resettlement of virtually the entire Lemko population (about 40%) from the so-called Łemkowszczyzny on the so-called Regained Territories. As a result of this process, Lemko families remaining in Poland, in Ukraine or in Russia, often underwent denationalization - polonization or Ukrainization. Despite many difficulties, the older generation tried to maintain their native culture and language. It became easier especially in Poland after the subsequent "thaws", when the Lemkos, starting from 1956, could apply for permission to return to their core areas - provided, however, that their lands and houses were not occupied by other Polish citizens who were settled. Areas of the Lemko region. (=> Wikipedia - http://en.wikipedia.org/wiki/%C5%81emkowie). We leave it without comment.

Poles lived in these areas during the war and remember it, so we consider the author's enunciations to be typical of desire. The mysterious shepherd Slavek and his two daughters could be just Slovak Ruthenians who dealt with carrying people across the border from Slovakia to the General Government and vice-versa. They could cooperate with the Polish and Slovak guerrillas, and - which cannot be ruled out - also Soviet. Let us not forget that the Nowy Sącz region was cut by courier routes leading from GG to Slovakia and Hungary, and from there to England.

Final remarks

I have an interesting observation - as I have already written here, Dr. Horak gave the coordinates of the Moon Cave - N 49 ° 02 '- E 020 ° 07' - which situate it near the village of Vikartovce in Low Tatras. Needless to say, it was there that two interesting things were discovered recently: an artefact from the old times called the Cyclops' Shield and just the uranium ore deposit. The question is therefore: did Dr. Horak know about this deposit? Dr. Horak must have known about him, if only because of his profession - he was a geologist. Perhaps he concealed his knowledge and did not pass this information on to the communists, because why? Then, in 1965, he decided to reveal this information in relation to his wartime adventures. So, we would have to deal with different artifacts in one relation? Why not? I am absolutely sure of one thing: Dr. Horak knew a lot more than he could say and wanted to convey this knowledge - which for various reasons was forbidden knowledge, so he could do it only by giving it a sensational story, which he published in a journal speleologists, but not only - by geologists, travelers and ufologists. And it did not disappoint - the matter gained publicity!

Why did he do that? It is simple - as a refugee from Czechoslovakia under the leadership of communists, former VIP and a Jew lived in exile with the communist death sentence, which Czechoslovak Št. B. and / or the Soviet KGB could have done it if it just fell into their hands. The secret he wore was burning him and he wanted to entrust it to someone. He did not have a family, and any attempt to contact the country had to end in detention and execution. That's why he wrote a sensational story and published it hoping that someone would read it and take it seriously. And that would be the solution to this secret??? - and why not? This reasoning has its weak point, namely that Dr. Horak published it under his own name. If he published it under a pseudonym, it would all be understandable. Well, unless he had effective protection against Czechoslovak and Soviet agents from the Fifth Directorate General of the KGB - called the Department of Wet Works. In short - the report by Ing. Hiltscher can be checked.

Interesting thing - like the cave in Zubrzym Wierch, this hamlet is not marked on all tourist maps. It is marked on military maps and Slovakian tourist and car maps. Oh, and one more thing - it is exactly opposite the Slovak Great and Little Sulin! Not only that - it is the easiest way to the top of Wubb Żubrac, where the entrance to the cave system is to be hidden, in which the mysterious Moon Shaft is supposed to be hidden. In fact, we paid attention to this fact being there in the summer of 2006, when we were collecting materials for our first book, as we wrote about it. Does this mean that it would be the "warm trail"?

Conclusions

On November 28, 2008, we went on reconnaissance to the area between Wierchmola Wielka and Żegiestów - Palenica in order to look more closely at the topography of the area, where the events described by Dr. Horak probably took place. What could we say? It was possible to determine that:

1. This area is quite difficult and covered. The border running in the mainstream of the Poprad river is relatively easy to cross ("break" - speaking in the language of border guards), which is why during the war and after it many smuggling and courier routes to/from Slovakia led through it.

2. The Poprad river counts in its widest point about 50 m and its transition in winter would not be too difficult, especially when it is freezing and low water levels.

3. On the Polish side of the border there are mountains with steep and forested slopes, covered with gullies formed by streams that can be reached on their slopes, which, however, would cause injured and tired people some or even serious difficulties.

4. The peaks of such mountains as Zubrzy Wierch and Pusta are almost invisible from the road leading from Piwniczna to Krynica. You can see them from the south only after entering the tourist routes leading from Żegiestów to Wielki Pusta and Jaworzyna - the way of Żegiestów Centrum - Pusta (black signs); trail from Żegiestów - Palenica na Pusta (yellow signs) and the route from Żegiestów Zdrój through Stawiska (759 m) and Trzy Kopce (687 m) to Pusta (blue signs).

5. The composition of stands of local forests agrees with those given by Ing. Hiltscher: pine, beech, birch and fir, sometimes spruce. There is no dwarf mountain pine, which was written by Dr. Horak.

6. There is no hamlet of Żdżaryka lying between Żegiestów and Żubrzyk. It could be this mysterious Yzdar/Zdiar from the accounts of Dr. Horak, in which the bail Slavek/Sławek lived with two daughters.

7. There is a stream called Zdziar (Ždiar), which flows from the western slope of the Bystre Mountain (807 m) and flows into Poprad

about 200 m south of the Wierchomla Wielka railway station. Maybe inż. Hiltscher found both Holes on this mountain.

8. The matter of the cave in the honor parts of Wierch Zubr is very interesting, because on some maps, it is visible about 50 m east of the top of this pillar, and on others, it is not present at all. It looks like Ing. Hiltscher did not look for her where the map indicated.

9. Similarly with the description of the cave in guidebooks - there are only references in them and there is no data about them.

10. Access to the possible underground of the western part of the Beskid Sądecki in the Pusta Wielka mountain group is difficult due to terrain conditions, which is why hiding the partisans was very possible, as they would have to absorb a large amount of forces and resources from the area, including the use of specialized SS units, Alpenkorpsu or Gebirgsjägern and other formations envisaged for police operations and fighting in the mountains.

11. The geological structure of these mountains is in sharp contradiction with what Dr. Horak writes in his report - these mountains consist of sandstone rocks of the Carpathian flysch - old Magura sandstones. The only possibility is a very deep tectonic fracture of the two-kilometer sandstone layer and the formation of a cave in the chalks of chalk carbonate rocks beneath them. This would be in agreement with what Dr Horak wrote, but so far no rock has been found beneath these mountains. Another thing is that no one was looking for it there.

12. In August 2009, we found one more interesting trail, and it is another hamlet of the village of Wierchomla Wielka, named even closer to the mysterious Yzdar - namely Izdwor! And not far from it, there is the massif of Pusta Wielka.

Intentions

Therefore, the possibility that the moon cave is highly probable in these Beskid Sądecki mountains is highly probable, nevertheless the possibility of a huge cave system under the group of Pusta Wielka is highly problematic, but it would be a mistake to exclude it. Therefore, we believe that further exploration should concentrate on the Polish side of the border. We intend to look for the Moonbelt in the area of Beskid Sądecki in the belt adjacent to the massifs of Żubr Wierch and Wielka Pusta. If there are traces of the presence of the Slovak sub-unit of the SNP fighters somewhere, just there. In addition, we will try to find information about the movements of the army in this area in the described period - at the turn of October and November 1944.

We cannot count on witnesses' accounts, because they must have already died out and did not transfer their knowledge to descendants, besides the villages in the vicinity of Żegiestów were first displaced from the Jewish population, which was liquidated in one of the extermination camps, and after the war the Lemkos were displaced - some of whom left to Ukraine, some to Polish Regained Territories and most probably to Slovakia. Perhaps we are right, and the Lunar shaft is somewhere in the mountains of the Polish-Slovak borderland.

CHAPTER IX

Reconnaissance in Yzdar

In August 2009, we decided to verify another trail in the matter of the Moon Cave. This is another key figure in our opinion from the accounts of Dr. Horák - she is Slavek or Sławek and her daughters - Anna and Olga. This time, we assumed that these people lived on the Polish side of the border - in Żegiestów or Wierchomla Wielka - in this context, and this mysterious "at Yzdar" from the journal Dr. Horák could simply mean "near the creek of Ždiar." So Slawek-Slavka should be traced just look for it in Wierchomla Wielka, because in our opinion, it was only there that he could live in. The Zdziaryki hamlet falls out because it was too close to the border and on the road from Muszyna to Krynica, where German patrols rode. While living in Wierchomla Wielka, he would not have a problem with that and he could go the simplest and most logical way to the Moon Cave and Zubr Wierch just along the stream of Ždiar.

However, soon another circumstance arose, namely that in Wierchomla Wielka there is also a stream and hamlet called Izdwor, which can also be read as Yzdar - pronouncing as Jzdiar! This stream flows through the center of Wierchomla Wielka. This hamlet is located just north of the center of this village, whose present-day central place is the church (formerly the Lemko church) dedicated to

Saint. Michael the Archangel. Thus, another place suspected of connections with the Moon Caves.

And to finish this topic - this place called Ždiar is located on the south-eastern slope of the Slovak Sliboň Mountain (789 m), just to the west of the village of Malý Lipník. So Slavek could live in this town. Another thing - if Slavek was a partisan, then the obvious thing had to be conspired. After the war, he could have been arrested - if he belonged to the Home Army, and either served his sentence or was shot as an English spy. If so, his data should be in the Nowy Sącz branch of the Institute of National Remembrance. Let us not forget that he was active in the areas of many partisan groups and could take part in the transfer of weapons, espionage materials and people across the border. Nowosądecczyzna was the main area through which courier routes ran - especially in the area of Eliaszówka and Obidza (Litmanová direction) and Tylicz and Muszyna (in the direction of Prešov) - therefore, courier routes had to run also through Wierchomla and Żegiestów.

Another possibility is his trip to Slovakia - which is probably assuming he was a Slovak. In this case, some trace should remain on the Slovak side of the border.

The third possibility is that he was deported with his family as part of the "Vistula" action to the Western Lands and lived there somewhere near Wrocław, Opole, Zielona Góra, Gorzów Wielkopolski, Szczecin or Koszalin or Słupsk.

The fourth possibility - it could be exported to Ukraine as part of the repatriation of the Lemkos, who were considered to be people of Ukrainian origin, and thus - here the trail is also lost, because we doubt whether there were any traces of Soviet control under him.

Surroundings of Żegiestów and Wierchomla Wielka after the war were displaced from the Lemkos and settled by Poles from the vicinity of Krynica. The Lemkos fell victim to the policy of "Ukrainianisation" conducted by Austro-Hungary in response to the alleged Muscular tendencies.The list of Lemkos, placed in Thalerhof, developed around 1930 by Father W. Kuryłła and probably incomplete mentions 1915 surnames. Among the 8,000 volunteers, only 30 Lemkos were counted, while the Second World War began when the Lemkos left for Soviet Ukraine and deeper into the USSR. Over time, these resettlements became increasingly intense (under the influence of Ukrainian nationalist propaganda) and It was not until the "Vistula" Action, however, that the Lemko population was accused of fostering the UPA rebellion in 1947, displacing practically all of the Lemko population (around 40%) from the so-called land areas. Le As a result of this process, Lemko families remaining in Poland, in Ukraine or in Russia, often underwent denationalization - polonization or Ukrainization. Despite many difficulties, the older generation tried to maintain their native culture and language. It became easier especially in Poland after subsequent "thaws", when the Lemkos, starting from 1956, could apply for permission to return to their core areas - provided, however, that their lands and houses were not occupied by other Polish citizens, who settlements of the Lemko region were settled (Wikipedia - http://en.wikipedia.org/wiki/%C5%81emkowie).

As for the population of Wierchomla Wielka, Wikipedia gives very interesting information about the fact that:

Wierchomla Wielka (Łem. Wierchomlia) was located on the Wallachian Law in 1595 by Cardinal Jerzy Radziwiłł. The founder was Mikuta Zubrzycki from Zubrzyk. The settlers were the Wallachian shepherds who came here. (So the Lemkos - so they were the autochthons here during World War II - author's note) They received as much as 30 years free to use. It was part of the Muszyński state. After the liquidation of the Muszyna state of Cracow bishops (1781), it was a good chamber (Kk Muszyner Cameral Verwaltung). In the 19th century, Poles and Jews also settled there. In the times of

Galician autonomy and until 1933, it was a single-brand municipality. Then she belonged to the collective commune in Piwniczna. The local population was repatriated in 1945 to the USSR (Ukraine). The remaining few families were displaced in 1947. The village was settled by Poles from the vicinity of Piwniczna. (Wikipedia - http://en.wikipedia.org/wiki/Wierchomla_Wielka).

Thus, the mysterious Slawek or Slavek could have been a Slovak as well as a Pole and a Jew. Let us not forget that the Cpt was also a Jew. Dr. Antonin Horák! However, there is a fundamental difficulty, because Germans during the war displaced all Jews from this area and put them in the Ghetto in Nowy Sącz, and then almost all of them were killed in the extermination camp in Bełżec.

And one more thing, this time regarding caves. In Wikipedia, we found an interesting mention of the caves under Pustka Wielka, and it sounds like this:

Wierchomla Mała is a village in Poland, in the Lesser Poland Voivodeship, in the Nowy Sącz poviat, in the Piwniczna-Zdrój municipality. In the years 1975-1998, the town was administratively part of the Nowy Sącz voivodship. The village is located in the upper part of the Wierchlomlanka stream valley, on the Tysina stream, at the foot of Pusta Wielka (1061 meters above sea level) and the southern slopes of Lembarczka (916 meters above sea level). This town was recorded in the Middle Ages, when Ścibor z Ściborzyc of the Ostoja coat of arms (1347-1414) in the autumn of 1410, on behalf of Sigismund the Luxembourger, invaded the Nowy Sącz land and burned and plundered Stary Sącz, he was repulsed in skirmishes in Piwniczna, Łomnica, near Wierchomla. The settlement was located in 1603 (according to other sources in 1601) on the privilege of Bishop Bernard Maciejowski. The founder of the village was Fedor from nearby Szczawnik. She was called Wierchomla Księżą because of the obligation to the priest Muszyński. The village was inhabited by the

Rusyns (Lemkos). It was a part of the episcopate of the Muszyna state. From 1770 in the Habsburgs and from 1781 it was owned by the imperial treasury (Kk Muszyner Cameral Verwaltung). The next years are common   happens with Wierchomla Wielka. Until 1947, the Lemkos were forcibly resettled to Ukraine and the Regained Territories. Wierchomla Mała is a tourist town in the Poprad Landscape Park between Muszynka and Piwniczna-Zdrój. There is a modern ski station (operating since the 1997/98 season) with 10 ski lifts, including the longest chairlift in Poland (1,600 m), also open in summer. Numerous regular sports events take place here - skiing, cycling, dog sled races.

Tourist attractions: old half-timbered cottages, numerous springs, caves under the Pusta Wielka Mountain, grills, paintball field, climbing rocks, hiking and biking trails, indoor swimming pool, rope park with a climbing wall, a hut on Wierchomla, near hostel Hala Łabowska, walking and cycling route to Jaworzyna Krynicka, from where you can take a gondola lift to Krynica.

In winter, skiers can travel through the system of ski lifts to Muszyna-Szczawnik. (Wikipedia-
http://pl.wikipedia.org/wiki/Wierchomla_Ma%C5%82a).

So there are some caves under Pusta Wielki! But this is the only mention of them - I have not heard any mention of them in any guide or on the Internet, but only in Wikipedia. Therefore, it should be checked in situ. And back to the well-known cave in Zubrzym Wierch, the analysis of the map of this part of Beskid Sądecki gives interesting results - as you can see in the pictures, we are dealing here with a certain phenomenon - we see that each edition of the cave in Zubrzym Wierch brings new location of the entrance to this cave, and so - on the map "Beskid Sadecki - Tourist Map", scale 1: 75.000, Warsaw 1974, this cave was not visible, which is not surprising, God's people's authorities were afraid that it will be used in evil targets by "enemies of the people" and "ugly, evil imperialists."

On the next tourist map "Beskid Sądecki - Tourist Map", scale 1: 75.000, Warsaw - Wrocław 2001, this cave is located in the Zubr Wierch highland side from the east.

Third map - "Beskid Sądecki", ed. III, scale 1: 50,000, Kraków 2006, locates the entrance to the cave on the western side of the Zubr Wierch peak, on the crest connecting this mountain with the neighboring Bystre Mountain (807 m). On the map from 2008 "Beskid Sądecki", scale 1: 50,000, Piwniczna 2008, the cave returns to the honor parties of Zubr Wierch from the east, and what's more - this entrance is located in a clearing, at least on a non-forest site.

Next, on the map "Krynica Zdrój, Muszyna and around", scale 1: 50,000, Piwniczna 2009 and "Poprad Landscape Park", scale 1: 60,000, Warsaw 2009, the entrance to the cave is on the western side of the Zubr Wierch Mountain.

In the interesting collective monograph "Atlas Mountains of Poland", ed. II, Warsaw 2008, this cave does not exist. It just does not exist anymore. There are, however, other interesting information that I will provide elsewhere.

What does this prove? It's just that either the tourist maps were made on the knee with accuracy that leaves much to be desired, or we deal with two caves in Zubr Wierch - one in an honored party from the east, the other on the ridge joining Zubrzy Wierch with Bystre from the west - in greater distance from the summit than the first one. And it seems also noted inż. Hiltscher.

There is another cave in this part of Beskid Sądecki, Devil's Cave Dziura next to the Devil's Stone located on the slope of Jaworzyna

Krynicka (1114 m) - but it cannot be taken into account due to its proximity to Krynica - a city with significant tourist and ski traffic - even in times of war. After the World War I, many buildings were renovated and several new ones were built, eg. New Mineral Bathrooms, the Lwigród guest house, and the New Spa House. At that time, a hostel was also built at Jaworzyna Krynicka and a cableway to Góra Parkowa, a winter stadium and a toboggan run. Apart from a purely spa face, Krynica-Zdrój was becoming a well-known center of winter sports at that time. It is enough to mention that there were, among others, European tobogganing and world hockey championships. The Second World War and the period of occupation interrupted the development of the spa. After the war, the resort was extended, modern sanatoriums, a natural medicine facility, a Main Pump Room with a concert hall, tennis courts and sports fields were created at that time.

(Portal Krynica24.pl -
http://www.krynica24.pl/pl/warto_wiedziec/krynica_historia/).

It follows that this area has been constantly penetrated by people and it was very difficult to hide there, so we can confidently exclude it from our considerations.

On August 28, 2009, we went on a reconnaissance to Wierchomla Wielka, where there are two streams with names similar to Yzdar. The first of them, Żdziar, falls from the Bystre ridge stretching between the peaks of Wierch Zubr and Rąbaniska (786 m). It was here that the team of Ing. Marco Hiltscher, who most likely reached Caves in Zubrzym Wierch. Of course, guerrillas might have been hiding there, and they were certainly hiding from crossing the border into the Slovak side. But it could not have been the Moon Cave.

So there are two places left: the first of them is the hamlet of Izdwor located on the stream of the same name flowing down from the slopes of Kiczera (840 m) and almost into the center of the village entering Wierchomlanka. We visit the old cemetery at the church adapted from the Orthodox Church. On it there are graves with characteristic 7-arm crosses with a diagonal bar and ordinary Latin crosses.

The inscriptions on the gravestones are in Polish and in Lemko. You can see the interpenetration of these two cultures.

The landscape is fabulous - we drive a car up the Izdwor creek to the house of village administrator Wierchomla. From here there is a magnificent view of the densely forested northern slopes of the entire Wielka Pusta and Zubr Wierch mountain nests. If Slavek actually lived here, he would not have the slightest problem getting to any place in the massifs of Pusta Wielka and Zubr.

We go further towards Wierchomla Mała. The valley is becoming more and more clench - on both sides there are quite steep slopes of mountains. The landscape begins to resemble the Low Tatras. The story of Dr. Horák and the location of the Moon-shaped Shaft in the Low Tatras comes to mind. Who knows if two older towns did not make a mistake for an elderly and controlled person? Their similarity is striking, which can be seen especially in the vicinity of the car park in front of the ski areas of Wierchomla Mała. We will come back here again, because in the meantime we went to Malý Lipník to see the third location with a name similar to Yzdar.

Indeed, from a quite scenic spot, we managed to photograph the floors, a wide slope called Ždiar sloping from the top of the Sliboň mountain towards the Poprad river and the border with Poland. If

Slavek lived under this slope, he could carry people on both sides of the border river and penetrate the caves on both sides of the border.

Unfortunately, it is not known whether there were some buildings in front of and during the war - now they are not there. The buildings visible in the pictures are the Polish settlement of Ługi. Only the buildings in the background and behind the river belong to Malý Lipník.

There is no building on the Slovak side on the road to Malego Sulín. The Poprad river is relatively narrow there and can be crossed without any minor problems. We are one sure - Slavek had to live in one of these towns. The lack of its traces on the Polish side is perfectly explainable just after the war. If Slavek was not deported from this area, he could go to a daughter who married one of Dr. Horák's comrades. That we can be sure of.

In the meantime, we decided to come back here and organize another trip to Pusta Wielka, where there is probably a solution to this mystery from 65 years ago. Studying the "Encyclopedia of the Polish Mountains" we find very interesting descriptions of the geological structure of this part of the Beskids, namely:

The Beskid Sądecki is formed by the rocks of the Magurian mire, that is the Carpathian flysch composed of sandstones, shale and marl. Within the Magurian Magura there are two subdivisions here: in the north of Nowy Sącz and put on it from the south - Krynica.

In the Krynica sub-unit there are conglomerates and coarse sandstones. In the mountains, there are sometimes outcrops of sandstones, eg. St. Kinga pod Skałką (1163 m), rocks in the "Wierchomla" reserve (which we are interested in, because they are the only outcrops in the area), Czarci Kamień and Skamieniała Daughter near Wierch under Kamień (1082 m).

In this region, there are also caves: Rysia (4 m long) and Niedźwiedzia (several hundred meters of corridors - and that means that it is possible to have large caves in sandstone massif!), several caves were discovered on the slopes of nearby Sokołowska Góra.[22]

Why have we emphasized these marls? Well, because marl is a sedimentary rock, usually gray; It consists of carbonates (calcium or magnesium) and clay minerals, used to make cement, also as a mineral fertilizer (artificial), has a weak, unpleasant odor. It reacts well with HCl (leaves a muddy spot). Margle are intermediate rocks between carbonate and limestone rocks. They are mainly made of calcite (from 50 to 70% according to Czermiński, from 33 to 67 according to Smulikowski), accompanied by smaller amounts of dolomite, siderite and clay minerals. They may also contain admixtures of crumb material, whose increased proportion leads to the formation of sandy or marly sandstones. Margle are generally less hard and compact than limestones, they also differ from them in a darker color. A characteristic feature of these rocks is a strong "demolition" of 10% HCl, during which a precipitate of clay minerals precipitates and remains. Recognizing them can sometimes be flat surface cleavage and even washed away, dirty hands.

(Margle -
http://www.mount.cad.pl/g/budowa/rodz_skal/prawa/index_skal/skaly/15.htm and Wikipedia - http://pl.wikipedia.org/wiki/Margiel).

And one more thing:

The geological structure of Beskid Sądecki, similarly to other ranges of the outer Carpathian Carpathians, is rather monotonous. There are quite thick (up to several km), a series of sandstones that are interwoven with shale, siltstones and claystones. These works are known under the common name of the Carpathian flysch and formed from the Upper Jura through the Chalk to the Lower Tertiary, or

---

[22] "Encyclopedia of Polish Mountains", Warsaw 2008, p. 309.

Paleogen at the bottom of the so-called reservoir. Tethys Ocean. During the Oligocene and Miocene tectonic movements, they were pricked out, i.e. detached from the original substrate, and pushed towards the north for a distance of several dozen kilometers to form a series of ponytailed overcoats. They rest on the crystalline bedrock of the Paleozoic platform. Beskid Sadecki mainly build fairly hard Magurian sandstones belonging to the Magurian mire. Next to them there are shale and conglomerate, besides, sub-Saharan layers, hieroglyphic layers, spotted shives and marls.

Smaller forms can be found in the "Baniska" reserve, in the Zadni Mountains (969 m) and the Wietrzne Dziurie. Interesting are outcrops of sandstone rocks at Wdżary near Prehyba, Kamień St. Kinga, Wierch nad Kamieniem or Diabelski Kamień near Jaworzyna Krynicka. Small caves occur on the slopes of Wierch nad Kamieniem (1064 m) and Wierch Zubrzy (860 m). Interesting wealth is mineral water, especially of Szczawa, Piwniczna, Żegiestów, Złock, Łomnica, Muszyna, Tylicz. The simplest of them, the bicarbonate-earth-alkaline oxalate are of infiltration origin, which means that the soaking rainwater is saturated with carbon dioxide. In this way they acquire aggressiveness by dissolving the minerals contained in the rocks. Proof of volcanic carbon dioxide underground is the presence of effusive rocks in Szczawnica and Slovakia.

In Złockie, Szczawnik, Krynica, Tylicz and Powroźnik, it uncovered the mofetic exhaust of carbon dioxide. Other, more complex, are: carbonate-bicarbonate-sodium or highly mineralized carbonated bicarbonate. They form as a result of mixing relic brines with infiltration seams (eg the Jan source in Szczawnica) or at great depths in the conditions of isolation from the surface of the earth and surface waters. The saturation of the relic brine is followed by volcanic carbon dioxide and the transformation of water into bicarbonate-sodium. (Beskid Sądecki - http://rzeznicy.republika.pl/t2001/beskids/bs04.html)

Is everything clear? Under the thick-layered sandstones of the Krynica mound there are, among others, marl magura coat. And this configuration can be just in the Wierchomla area. In addition, the poisonous fumes that Dr Horák mentioned in his diary could only take place there!

So we decided to go to the area of Wielka Pusta to explore its secrets.

CHAPTER X

Courier routes

Sorrels are mineral waters containing more than 1 g of free carbon dioxide (CO_2) in 1 liter of water. Sorrels are infiltration waters which, when they soak in the ground, encounter carbon dioxide fumes (in Poland related to the volcanism of the Carpathians in the Trzeciorzędzie). By saturating it, they become more chemically active and dissolve the rocks in which they flow (mineralize). They occur in the southern part of the Sądecki Beskids and on the eastern border of the Pieniny, in Szczawnica, Krościenko on the Dunajec, Krynica, Piwniczna, Muszyna, Wysowa, Żegiestów-Zdrój, also in Szczawno-Zdrój. Carbon dioxide is of volcanic origin.

Chemically, it is a concentrated carbonic acid with admixtures of mineral salts. As you can see, there are many mineral springs in the vicinity of Wielka Pusta, and therefore there may also be rock spaces filled partially or even completely with carbon dioxide. Actually for sure.

We would like to draw the Reader's attention to a very interesting aspect of the matter of the Moon Caves and the attack on Gen. Władysław Sikorski in Gibraltar.

"Where is the river and where is the forest?" someone will say. Yet there is something to do with these two events. First things first. In the third part of the trilogy about the attack on Gibraltar "Gibraltar and Katyn", its author - Tadeusz A. Kisielewski[23] writes, among others about courier routes in the Nowy Sącz area during World War II.

Why is the matter of courier routes so important? Because of the fact that Cpt. Horak and his men somehow had to get to Poland - that is, the then General Government - where they were kept in the Moon Cave. In this context, Slavek could be someone from the Polish or Slovak conspiracy specialized in smuggling people across the border between GG and Slovakia. Such people usually have a few meteors (melins) and one of such melins could have been the Moon Cave.

In the section of the border described by us, according to Tadeusz Kisielewski's documents, there were at least four routes or metastases for couriers from GG to Slovakia and further to Hungary. He cites the opinion of Jan Cieślak about metastatic channels in Nowosądecczyzna:

Powiat Nowy Sącz:

Metastasis used for both the military and the departing Jewish population led from Krynica through Tylicz to the town of Lenartec[24]. The second route led from Szczawnica to Krościenka through Dunajec to Orlov in Slovakia and further to Košice. Around 8.000 people passed through the Nowy Sącz poviat during the occupation.

Limanowa Poviat:

---

[23] Ed. Rebis, Poznań 2009.

[24] It's about the village of Lenartovo.

Metastasis from Limanowa went through the roads to Krynica - Musziny and Orlovo[25] on the Slovak side and Żegiestów - Alverna[26] in Slovakia.

Jan Cieślak also provides interesting information about border traffic in these areas, and so there was also such a reverse movement across the border (ie from Slovakia to Poland / GG) and was tolerated by the Germans.[27]

It also lists other metastatic routes that were most commonly used:

Variant:

a) Nowy Sącz - Rytro - Komarzyska - Eliaszówka - Lubonia or Kežmarok - Prešov - Košice - Budapest;

b) Nowy Sącz - Bardejov - Salgotarian - Sarospartak - Budapest;

c) Warsaw - Nowy Sącz - Szczawnica - Poprad - Dobšin - Rozsnyo - Budapest;

Where the border was crossed:

In variant A - through the top of Eliaszówka and Koszycki Las;

In option B, directly to the Slovak village of Lenartovo, and to Hungary and the village of Garany;

In option C from Szczawnica to the village of Leśnica, and to Hungary through Park Andrassych nad Velko Polono to Rožňava.

I had other passages, for example Sanok - Uzhgorod or Zakopane - High Tatras - Poprad[28].

---

[25] Town of Orlov near Penava.

[26] Unfortunately, I was unable to determine what the town was and where it was located. Maybe it's about some sort of thing a town located in Hungary's part of Slovakia or Hungary, which was renamed after the war.

[27] Cieślak's Papers, Stanisława Groblewska, typescript, pp. 1 and 2.

[28] Ibidem.

We found the next trail on the website of the "Internet Guide to Poland", where under the slogan "Żegiestów Zdrój" we found the following passus:

During World War II, Żegiestów was one of the courier centers organizing metastases to Hungary. During the war, the spa did not suffer, thanks to which it could develop without major obstacles[29].

Another trace indicates Józef Bieniek in the article titled "Sądeczanie na border route" from the website "Once upon a time in Nowy Sącz" - http://www.nsi.pl/almanach/art-ludzie/sadeczanie_na_kurierskim_szlaku.htm :

The border location of the Nowy Sącz region and its topographic specificity imposed additional tasks in the local lands during the Second World War, closely related to the anti-Hitler resistance movement. They were the transfer of Polish people to the Polish army in the West across the Polish-Slovak border as well as courier, that is, the communication service on the lines binding the national underground authorities with the government and the Supreme Commander in exile.

It began on September 17, 1939, when - as a result of the agreement of the Polish authorities with the governments of Hungary and Romania - borders were opened and the parts of our southern neighbors were joined by the armed forces, and with them the supreme command and authorities of the Republic of Poland and thousands of people from the interwar political leaders, social and cultural. It is the transition of the leading spheres of Polish pre-war Poland to the foreign and gradual over time their departure from their national borders, with the simultaneous formation of the national underground, created in the first place the need to establish a permanent communications network that would link the Polish

[29] "Internet Guide to Poland".

authorities' dispatching centers with the leadership of the national resistance movement .

The problem of building a communications bridge across borders and states was already dealt with in October 1939 by special departments at the Polish government in Paris and the General Headquarters of SZP ZWZ in Warsaw. , bases and auxiliary facilities in Budapest, Bucharest and Belgrade.

The main barricades on the "bridge" routes were interstate borders, except that the specificity of the war situation excluded the eastern and partly northern borders from the operational capabilities, shifting the main burden of the problem of external communication to the west and south.

The communication trend, taking place only on the conspiratorial planes, hence deprived of openness of action, legal grounds and protection from official legislation, could not exist and act in a social vacuum. Wanting to exist and operate in designated directions, he had to create his own organizational forms and a whole range of special links, with a well-developed background and auxiliary services. I am talking of conventional communication, that is, land-sea courier, which, especially in the initial period of the war, was the only form of binding the national underground with the free world. It was only later that radio communication was launched, and from February 15, 1941, air force came to courier services and radio stations, which from British and later Italian bases made flights with airdrops of people, mail, money and weapons for the resistance movement in Poland. The courier, however, existed and operated until the end of the war, performing tasks that neither the radio nor the discharges were able to accomplish.

So about courier. Or rather about those parts of it where the Judenats, the leading courier frontiers, operated on all positions of the Military Base of Transfer and Communications No. 1, codename:

"Romek", "Liszt" and "Pestka" and the governmental "W" facility during all the years of the war. in Budapest. They were Jan Freisler ps. "Sądecki", Franciszek Krzyżak aka "Karol", Leopold Kwiatkowski ps. "Tomek", Zbigniew Ryś aka "Zbyszek" and Roman Stramka ps. "Bardyjowski" and others, all of them had a few pseudonyms and the same number of "left" surnames ", changed every now and then, when they became too familiar, or when there was a scandal in the circle of" green borders ".

A specific drama for those who write about these matters is the fact that courier is the subject of an unbelievable wealth of threads, issues and events that cannot be treated more widely in a calendar. We will show only the most important fragments of the larger whole.

Being the first or one of the first, she started working on the Sądecki episode of communication and communication, organized on behalf of the Warsaw cell of the Union of Reserve Officers (ZOR) by a judge, foster child of the Zenglów family (38 Zygmuntowska Street), cf. Klemens Konstantego Gucwa ps. "Mountaineer".

"Góral", a friend of "Kostek", a foster-boy of Military Railroads, very popular and well-liked, was actually the creator of the courier team in Sądecczyzna, who worked from March 1940 to the end of the war, inscribed in golden letters to the history of foreign communications. The Góral team started with metastases, the first clients were officers sent by the headquarters of ZOR in Warsaw. After time, the list of refugees widened all over Poland. "All those who managed to get in touch with him went to Góral".

"Góral" completed an excellent team composed mostly of railway sons, former colleagues from the Railway Approach.

Military, who were in love with the Beskidy "zbyrki" rallies, great skiers and athletes, the first who went to the border service were: Jan Freisler, Leopold Kwiatkowski, Franciszek Krzyżak and Roman Stramka, they were also the first to pass the "live merchandise" in the

Polish attaché at Vaciutca, "caught the eye" of officers directing the affairs of refugees and were engaged on courier posts on the Budapest - Warsaw route. On April 1, 1940, after taking oaths on the ZWZ rotary round, all four commenced the communications service on the basis of the Military Base of Transfer and Communication No. 1, codename "Romek", in Budapest.

The courier service proceeded according to a pre-arranged schedule. The designated day and hour, at a set point, the courier received the mail encrypted in the form of microfilms and other secret documents, usually in such a small form that they were hidden in the pen holder, a toothpaste tube, a compact or a suitcase handle - and he moved Let's go. First, take the train towards the Slovak border. One or two stops before the border got out and knowing the area, he waited for the dusk to dusk. At night he crossed the Hungarian-Slovak border, and then he reached the Polish border near the Polish border, where he waited again for the night under which he "skipped" the border and landed on his native land for a Slovak or Grenzschutz patrol. Because then either he died in the fight or Auschwitz devoured him.

Of course, the above course of the rally is very simplified and schematic, in fact it happened differently: hundreds of obstacles and unforeseen events arose on the courier route, only those who had the maximum amount of physical and psychic strength, adequate mortar, cold blood, cunning and courage, or those to whom he grinned at a critical moment of gracious fate.

But let's come back to our Judiciary again and follow their fate. The main points of the courier's oath submitted to the Military Base were not to contact any organizations that communicate with the country, keep secret service, do not accept any orders from institutions and persons, except for the Romek base mail. absolute conspiracy and discipline and security of people and affairs, which in

Budapest, full of intelligence agencies of all states interested in war, was an absolute necessity. "All these conditions and reservations did not reach the conviction of the four of us. to all and everything that was called: Poland - Fatherland, being extremely friendly and friendly to the whole world, they were not able to refuse the requests of various activists of the Polonia and, taking a voyage to the country, took their mail, with the obligation to pay plane at the indicated address. Caught by counterintelligence for this kind of "foreign services", were released from Base command and Commander in Chief of the AGM, gen. Grota - Rowecki dated 31.10.1940 r. handed over to the disposal was founded in Budapest institutions 'W' (Hungary), which is a "pillar" for the "bridge" of civil communications between the authorities in exile and the Government Delegacy in Warsaw.

The courier, in spite of appearances of calmly overcoming designated and organized routes, was in fact a kind of front on which the arrows at night and in the sides of the borderland blasted and the soldiers of communications or people towed by them were falling into enemy hands. There were many such events. But here we will only mention a few more serious ones. In February 1940, the courier of the General Headquarters of the ZWZ, Krystyna Michalska, codename "Michcik" entered the "Góral" route. Her phased point in Nowy Sącz was the Harsdorf house cooperating with the "Góral" grid at 78 Batorego Street. One of the rallies was in MichaĹ, in May 1940. On May 23, she found herself at the Harsdorf site and here she was followed by the Gestapo Arrested, she died shot near Tarnów. In the meantime, a number of Warsaw organizations that served "Góral", merged, creating so-called. Central Committee of Independence Organizations (CKON), with Ing. Ryszard Świętochowski at the forefront. Świętochowski, a loud politician from the Paderewski group and General Sikorski (Front Morges), took a decidedly hostile attitude towards the official military formation, which was the Union of Armed Struggle. So he fell into conflict with his closest friend, General Sikorski, and, to clarify the situation, he went on a trip to Paris. A wealthy man, accustomed to luxury driving conditions in international coffee machines, and typical townspeople, went for a long way for a walk in Aleje Ujazdowskie, in an elegant suit and in varnishes. So, when he appeared in the den of "Góral" with a request and help on the way to Paris, "Góral" decided: "Well, sir, it's my duty to help you, but I warn you that in your age, in the absence of a sports mortar it will not be an easy matter. We will not understand you for years, we will not learn the mountain steeples, but we can and will dress you in the right suit. "In the first half of May, Świętochowski, disguised as a tourist and devoted to" better of the best", Jan Freisler, pseudonym From Nowy Sącz, a taxi to Jazowska Obidza, where two great Szczawnica highlanders (the Mastalscy brothers) were waiting to bring Świętochowski to the first stage point at the border, but after a few kilometers, somewhere in the vicinity of Przystopia, Świętochowski fell he said: "Take it a step further, I cannot do it!" A stretcher was carried from a branch to Szczawnica, and on the way he caught pneumonia, and there was no way to go along with the sick doctors: Zdzisław Kołączkowski, a sworn physician and a certified nurse (on the ideal posture of resistance) Melania Czamara aka Jedlin". After two weeks, when Świętochowski felt better, Freisler organized the engineer's transfer to the village of Leśnica lying on the Slovak side. In Leśnica an appointment of a taxi driver, Józef Lach from Poprad, who carried Świętochowski with Freisler and a guard to the village of Mala Veska, located just below the forested hills, the top of which ran between Slovakia and Hungary. At dusk, they were on their way. Unfortunately, fate once again crossed the intentions of the engineer. At one point, they stumbled upon a patrol of Slovak "limbs" who shouted: "Stojte! Ruki hore!" - They started running towards them. The protection of the engineer - green border dodgers - disappeared into the forest thicket. Freisler, carrying Świętochowski's bags with a pile of money and an important post for General Sikorski, did not have much to wait for. The lonely engineer tried to escape, but he got tangled up in the roots and fell, getting into the hands of the Slovaks. The next stage of his fate - the Gestapo prison in Nowy Sącz and the departure of 11 September 1940 to Auschwitz. From that moment, no trace of Świętochowski has been lost. There was only the fact that Freisler received Sikorski's high distinction and praise for saving Świętochowski's post.

We are still stuck in 1940, because just this year on the courier routes in Nowy Sacz one more peculiar fact occurred: the arrest and rejection of the emissary Jan Kozielewski, vel Piasecki, aka Karski from the hospital in Nowy Sącz. In addition to the already mentioned lines, there was a transfer and communication route within the newly established ZWZ region, called the Inspectorate of Nowy Sącz, code name "Sarna". Ryś, after returning from the September front, dealt with the transfer of friends of officers and colleagues. It had its own route via Barcice - Prehyba - Szczawnica -Leśnica. In January 1940, he joined the ranks of the ZWZ and gave his tour in the service of the Inspectorate, whose head, Major Franciszek Żak, pseudonym. "Franek" or "Siwosz", commissioned Rysi to organize and command the Department of Transit and Communication Protection. One day, in the last decade of May 1940, to the house of Rysia at ul. Matejki 2, the liaison man brought a young man who introduced himself as Jan Piasecki and, providing the password set for "Zbyszko", he asked for help on the way to Budapest, where Piasecki was located in the house of Mr. Żaroffe at 13 Lwowska Street and communicated with Major Żak, he entrusted the petitioner to the best of the guides, Franciszek Musiał aka "Myszka" from Piwniczna, who already had 33 "trouble-free" courses on the Nowy Sącz - Budapest route. Thirty-fourth ended in drama.

They left in the evening of June 12. He ran Musiał with a colleague, Władysław Gardon, who insured. Late at night they reached the Hungarian border, to the village of Demjata, where Franciszek Muszyński's "Myszka" had his resting point before the "jump" across the border. He did not know that his loyal melinarz had in the meantime been taken over by the service of the Slovak gendarmerie as a "catcher" of refugees, much more than art. He adopted the "Mouse" band as if warmly, hosted, put to sleep and disappeared. He returned in half an hour, but with a group of gendarmes. Piasecki had only enough time to destroy the films he carried with photographs of executions carried out by the Gestapo on the Polish population. The arrested were transported to the detention center in Kapusany, where Piasecki, fearing torture, pulled out a razor blade from the shoebox and cut the veins of both hands. The gendarmerie appeared on the prisoners' alert and, having put on temporary makeshift dressings, she took the wounded to hospital in Nowy Sącz the same night, notifying the Gestapo, who put two navy policemen at Piasecki. The emissary was taken care of by a physician secretly in the inspector's sanitary unit, Jan Słowikowski , who first notified the commandant of "Sarna" about Piasecki's fate .Mjr Żak immediately took up the emissary's fate and instructed Rysi to organize the wounded reflection as fast as possible before they transport him The more so because the military and civil authorities of the District in Kraków were interested in the matter, as the emissary was a highly noted figure in the diplomatic world and he was very important to Paris for very important arrangements from the emerging national underground authorities, for the emigre government in Angers. the efforts of dr. Słowikowski and Rysia, composed of: Ryś, lieutenant Karol Głód , Tadeusz Szafran , Józef Jennet and liaison officer Zofia Rysiówna Piasecki was abducted at night to Marcinkowice, where under the protection of Jan Morawski and Felix Widła he recovered. case ca. We will limit ourselves only to the mention that for many post-war years Piasecki, under the well-known name Jan Karski, was a lecturer at the Georgetown University in Washington. In 1990, we listened to him on television.

However, when we mentioned Rysiek, we must show him in a broader aspect, as one of the most outstanding soldiers of Budapest's "Romek" base, later codenames "Liszt" and "Pestka." Sądeczanin, born in 1914, completed the two years of law studies and the school until the outbreak of war. After the described action to release Piasecki, threatened with arrest, he went to the sisters living in Warsaw, addressing the disposal of the General Headquarters of the ZWZ, directed to the Foreign Communications Unit, underwent several months training in the field of rules courier service and languages: Slovakian and Hungarian, after which he was transferred to Budapest as a courier of the Military Base "Romek", where he worked until the end of the war. In the final phase he was the head of communications for the country, and from April 1944, when the Germans occupied Budapest, and the Base under the name "Pestka" moved to Bratislava, Ryś took the post of deputy commander of "Pestka", Colonel Franciszek Matuszczak pseudonym He was very active, he made 108 rallies on various sections of the line, mainly on the route called "Tavern": Budapest - Roznawa - Slovakia, to the border points of "Karczmy" at Dominik Staszek 's Łapsze Wyżne, or at Pajorów in Dursztyn.

Arrested in April 1945 by the NKVD in Bratislava, he escaped transport to Siberia, but he spent several years. Released, he completed his studies in Wrocław, where from 1959 he ran a lawyer's office. He died on September 29, 1990 in Wroclaw. Behind the coffin, a number of medals were carried, including the Cross of Valor and Virtuti Militari, three times awarded. And the army company said goodbye to Lt. Rysia with a salvo of honor. Soldier.

The year 1941 marked a loss for the courier: on March 18, he died from the bullets of the Slovak border guards, going to Budapest, the head of the "Świętochowszczaków" line, cf. Gucwa "Góral", left name Adam Opyrchał . Seriously wounded and brought up by colleagues from the square of the skirmish on the Hungarian side, he died the following day in the Košice hospital.

His place in the team was taken by Jan Freisler, who in turn was subject to an employee of the "W" facility, with the function of communication manager for the country, Mr. Wacław Felczak aka "Madziar." The Freisler team was a team only at a glass of wine. the team was an individualist with individual tasks and walking, like a Kipling cat, by his own roads Freisler belonged to the elite and was intended for special tasks, Stramka - an exceptional individualist and skitter, operated by the People's Party, Teutonic - pupil of OMTUR, he worked for the PPS. different orders, but it must be dealt with separately.

Of the many route variants that the Sądeczanie used to migrate, three eventually cleared up:

1) Nowy Sącz - Krynica - Tylicz - Muszynka - Bardejov - Košice - Budapest,

2) Nowy Sącz - Kosarzyska - Sucha Dolina - Eliaszówka - Jarabina - Keżmarok - Budimir - Košice - Budapest,

3) Nowy Sącz - Szczawnica - Poprad - Mala Veska - Roznava - Budapest.

The three routes on the Slovak side were served by the Romek base and the "W" taxi drivers: Józef Lach from Poprad, Otto Ludwigh from Kieżmark and Istvan Burger from Dobsina. In 1941, the communication traffic on the routes mentioned above became very thick and exceeded the safety standards. In order to avoid mishaps and feed, the head of communications to the country in the "W" establishment, W. Felczak, ordered Kwiatkowski to organize a new route through Orava. Kwiatkowski was accommodated in Rabna Wyżna with his aunt Katarzyna Kwiatkowska and based on the house of Sabina and Jan Obertaczów in Orawka , lying on the Slovak side at the time - launched a new route: Raba Wyżna - Orawka - Kežmarok - Košice - Budapest It had no official name, and in common language it was called "Orawa". Mainly Poles worked on it from the areas connected to Slovakia and their friends, the Slovaks on the pro-Polish orientation. The courier on the Polish section Kraków - Warsaw was a well-known musicologist, professor of the conservatory, having Swiss citizenship, Konstanty Regame . "Orawa" worked for liberation, with a small break during the Warsaw and Slovak uprisings, when Kwiatkowski settled in his hometown after various perturbations, where he died on February 11, 1968.

And others? So the "leader" Freisler left the border front in March 1944, after the Germans occupied Budapest and the bloody trial of the Gestapo with more prominent Polish activists in Hungary, he descended from one front to enter the other. In June 1944 he organized the germs of a partisan unit "Świerk", with which in August it joined the branch of Lieutenant Julian Zubek, pseudonym In command of the platoon, he did a lot of successful actions and in January 1945 he left the detachment and revealed himself to the Security Office. After serving his sentence, he vegetated on the social margin, constantly under surveillance and harassment - he died in Warsaw on 10.10.1964.

In April 1944, the Gestapo arrested Stramka and Lenza at the station in Budapest. They were sentenced to the concentration camp at Gusen. But only Lenz reached his destination. Stramka, who had escaped from prison four times, this time also managed to escape from transport and return to Budapest, where, hiding with friends of the Hungarians, he lived to see freedom. After the war, he returned to the country and settled down in sport and tourism. Even as he arranged himself and began to legalize restless life. And here he, who so many times on the courier route escaped from the cold deaths, died in the stupidest way in a motorcycle accident on 1/9/1965.

Zamość Franciszek Krzyżak landed the most miserably from post-bourgeois trouble. When in the autumn of 1942 the whole activist of the underground PPS, called Freedom - Equality - Independence (WRN) was arrested, Krzyzak was the leading activist in the pre-war years of the PPS and it was ordered to transfer from the courier to a political work and rebuild the Nowy Sącz underground.

The courier did not completely abandon the mission, but he fulfilled his organizational tasks quickly, regenerating the broken leadership of WRN, in which he assumed the function of the Commander of the People's Guards, In spring 1943, when the Council for Aid to Jews was called "Żegota", Krzyzak became its member and The same line operated by the brothers Władysław and Kazimierz Świerczek, went to the free world of over 50 people. The same year, the Teutonic Knights organized a division of the People's Guard, which, leading to liberation, carried out a series of sabotage and subversive actions After liberation, for a time he searched for his place under the sky of People's Poland, but he had no special resistance to the new system since he was young, so he joined the current of political life, graduated with a civil engineer diploma. Permanent in Ta where, for many years, he worked as the director of the Municipal Repair and Construction Company. He died in the eighties.

And so, as you can see from this text - the border was very "hot" and people ran through it in both directions. Poles, Slovaks, Hungarians, Romanians and not only worked with each other here. Another interesting - and very much! - the traces are the memories written by Seweryn A. Wisłocki in the article entitled "Kurierskie szlaki przez Łemkowszczyznę" posted on the website of the Beskid Niski local government –

http://www.beskid-niski.pl/index.php?pos=/lemkowie/historia/kurier, and which sounds literally like this:

In the consciousness of the average Pole, the conviction persisted that courier communication between the command of the underground in Poland and the government agencies in exile in Budapest or Bucharest was kept exclusively by the Tatras. This is the effect of literary, widely available descriptions of courier adventure, promoting in the collective consciousness the nation of the heroic and tragic vicissitudes of Helena Marusarzowna, Bronisław Czech and Stanisław Marusarz, later called the "Grandfather".

This is only part of the truth and the image of metastatic and courier routes should be extended in Polish society and this is extended to include knowledge about the Polish underground in the border zone of the Sądeczczyzna and the Low Beskids, including the participation of the Lemkos in it. Practically, on this subject, one man wrote competently and honestly. It was Józef Bieniek from Nowy Sącz, author of source, documentary and publication. Particularly noteworthy is the work published in "Rocznik Sądecki t.9", Nowy Sącz, 1968, entitled: "Between Warsaw and Budapest (About Nowy Sacz during the occupation)."

"Rocznik Sądecki" is a publication of the Branch of the Polish Historical Association in Nowy Sącz. At that time, he had the circulation of 1 thousand copies. Who reads specialist publications, and what are his strengths at such a low circulation? The communist authorities promoted what was necessary or convenient for some reason and did not allow for a wider circulation of information that could compromise the policy of the PRL in the Lemko region.

Just think what it looks like in the light of the facts of the fight against the Germans and their ranks against the Germans, their cooperation in organizing and maintaining courier and metastatic routes to Budapest - the basic motivation of the "Vistula" Action about the Lemko's generally hostile attitude to Poland, Poles and exclusive support for UPA branches ?!

In general, as well as other historical painful matters, it is not allowed to generalize, generalize or apply the principle of collective responsibility.

We must slowly and stubbornly sow the wheat from the chaff, so that historical justice will be done, as Józef Bieniek rightly believed. He is also the author of an important publication "Lemkos in the service of the Polish Underground" published in "Tygodnik Powszechny" No. 15 of 14 V 1985. After its appearance, I made contact by letter and later personal with Mr. Bieńk and gave him supplements and I noticed minor errors, such as turning (for lack of other source materials) my grandfather's name: Harasym Gromościak - instead of Harasym Gromosiak I know that he received all these remarks and materials very warmly, nevertheless, being ill and elderly (he was 74 years old), he could not publish them anymore, so based on the abbreviation Bieńk's publication, as a basic sketch, I would like to present information obtained from Basil Owl , an outstanding activist from Krynica.

(The Sow family from Łabowa had been friends since the time of the pre-war with our family, ie Hromosiaków in Krynica).

At the beginning, I will quote the most important fragments of Bieńk's publication, and then there will be time for supplementation.

"In the area covered by the present memoirs, the Lemkos constituted an ethnic village, sharing a compact enclave of Polish settlements from the Slovakian border, from Szlachtowa in Nowotarskie, with a small break in the area of Piwniczna, to the border of the Jasło poviat they occupied over one hundred villages. All roads led through the Lemko area - there was no question of wider transfer activity without the involvement of the local population.

This attitude of West Lemkos became for ZWZ-AK a major opportunity for action in its frontier operations.

Refugees choosing the Nowy Sącz direction landed primarily in Nowy Sącz, from where the activists of metastatic agencies directed them to the appropriate addresses in Muszyna, Tylicz and Krynica, the latter playing a major role in the metastases handled by the Lemkos.

Such points were located in private holiday homes, the owners of which maintained friendly relations with the Lemkos in nearby villages, due to the supply of agricultural and forest produce.[30]

So when in the autumn of 1939 the other patients and holidaymakers came to ask for help in crossing the border - the issue was the simplest in the world, and the exit one - trusted Lemkos. Especially those who struggled with smuggling. For they knew best the secret gates in the border wall, the customs of border guards; they also had relatives or friends on the Slovak side. Which, of course, was of paramount importance for the transfer. It was on such principles that dozens of houses in various border towns were doing their job.

Lemkos were known primarily for their absolute integrity. In addition, they were characterized by a rare straightforwardness, raw traditionalism, great immediacy and cordial hospitality. These virtues, combined with a great knowledge of the terrain on both sides of the border - have created in some circles of refugees a certain conviction that whoever gives his fate to the guide Lemko - will achieve the intended goal with absolute certainty.

---

[30] Our emphasis.

Such convictions grew with the Hungarian news from the inflow to the country, from those who, with the help of the Lemkos, reached the first stage of the wandering road - Budapest, and who, in their wake, recommended Lemko services.

Because some of the Polish guides, wanting to get customers to transfer (paid) impersonated the Lemkos - the refugees established a way to help them get to know the flawlessly well-placed guide is Lemko. Such a way was a kind of examination of the knowledge of a prayer. Of course, in Lemko language and according to the Greek Catholic ritual, while the very moment of cross-examination has already explained the matter in a reliable manner.

In Sądeczczyźnie, Polish-Lemko cooperation began in Grybów, where metastases were carried out by the ZWZ branch there, served by two Lemkos, brothers Grzegorz and Józef Wilczaccy from Florynka . The musketeer was the most active in helping refugees. The village was wedged with a wedge between a team of Slovak settlements, inhabited entirely by the Lemkos - it had ideal conditions for crossing the border. This is the place where the largest number of refugees directed by Krynica's meters went : Zofia Sas-Bojatska, Helena Witkowska, Michalina Piszowa, Stanisław Kmietowicz, Jan Tryszczyła and others. In Muszynka she helped refugees almost the whole village, and the role of guides was made up of brothers Izydor and Jan Cieniawscy , brothers Daniło and Teodor Kostyszak , Wasyl Łychański, Teofil Chowaniec, Włodzimierz Nesterek and others. In Tylicz, the functions of the guides were, among others, Andrzej Garbera, Jan Koczański , Teodor Dutka and Anastazja Pawliszak, and brothers Stefan and Włodzimierz Rystweej .

They served the local transfer facility headed by Kazimierz Janiec and a number of transfer points in Krynica. It was in Tylicz that the scale of the Lemko service for the Polish Underground took on an extremely large character, which was due to several reasons, including benevolent attitude, for the Poles, the head of the Tylicz commune, Tymoteusz Rybiński and the parish priest of the Greek Catholic parish of Fr. Włodzimierz Mochnacki. The Lemko folk of Powroźnik also showed a highly positive attitude towards Polish refugees. This village, lying between two health resorts - Muszynka and Krynica - had the everyday opportunity to interact with the Polish element, which meant that during the occupation, the local population stood on the Polish side. This attitude was paid for by a number of victims. The following people died in Oświęcim: Józef Bartosz, Dymitr Galik, Konstanty Galik, Wasyl Halczak, Marek Kapuściński, Ludwik Smoczyński, Grzegorz Węgrzynowicz and his brother Władysław . They returned from prisons and camps: Mikołaj Halczak and Józef Pańczak.

In Powroźnik, two metastatic streams coincided: Muszyna and Krynica. Here, in dozens of Lemko houses, refugees gathered and waited for a moment of "jump" through the nearby border. Hence the Lemko guides took them along the Wojkowa - Dubne - Obrucne - Lenartów route to the railway station in Bardejov.

Among the many Lemkos from Przroźnica, the greatest merits of the Polish cause were given by Jan Galik and Jan Peregryn, whose house stood off the beaten track, at Szczawniczy Stream, was one big "den" for refugees directed here by Muszyna's transfer points. In the Muszyna-Powroźnica network also worked as guides Lemkos from neighboring villages: Maksym Kieblisz and Aleksander Lelito from Wojkowa, as well as Filip Polaczek and Adam Pyda from Dubne.

Harasym Gromosiak from Krynica Wieś, who was arrested in January 1940 for holding Polish officers abroad in his home, cooperated with this network - he suffered more harassment. Finally, losing sight of the beating - he was shot in a Krakow prison at Montelupich street.

Outside the courthouse, Lemko's service for metastases was recorded in the Gorlice and Jasło poviats. Two grids worked in Gorlice: a joint Groblewski from Bystrzyca Szymbarska and Zgórniaków from Nowodworza, and a "scout" led by a teacher and scoutmaster Maria Rydarowska . In both, the Lemko guides played a very important role. It was similar on the Jasło route, codename "Corridor", which led through the Dukielska Gorge and operated by the Lemko team, brought out into the world a few hundred future soldiers of General Sikorski.

Great knowledge of the area and people, and the great discretion with which the Lemkos treated the problem of metastases - she kept them from pouring in and betraying. The discharges here were much less frequent than in the regions where the metastases were exclusively based on the Polish element. The main reason for the loss of Lemko were the refugees themselves, who arrested in Slovakia and handed over to the Gestapo, did not always manage to withstand the torture of the investigation and broke down, revealing the addresses of the Lemk guides.

In addition to the aforementioned Gromosiak, they died in prisons and camps for participation in the underground: Antoni Demczak and MD Stefan Durkot from Nowy Sącz, Aleksander Hnatyszak from Grybów, Wasyl Dubec, Grzegorz Klucznik, Jan and Grzegorz Wilczacki from Florynka , Wasyl Porucznik and Piotr Rydzanicz from Mochnaczki Niżnej, Jan Peregryn from Powroźnik, doctor of law Orest Hnatyszak together with his son Igor from Krynica, Teodor Rusyniak from Wierchomla Wielka and Danyło Kostyszak, Teodor Kostyszak, Jan Cieniawski, Izydor Cieniawski and Wasyl Łychański from Muszynka. You must also include the list of Lemko victims Krynica student, Igor Trochanowicz , who was arrested for conducting a group of Poles in November 1939 - three years he was in prisons and camps.

These are, of course, fragments of a large whole, the elaboration of which is very difficult due to the lack of source materials and the scattering of the Lemkos in various regions of Ukraine and the lands on the Odra and Nysa. But wherever they are - let them be convinced that their heroic attitude on the common front with the Poles, the battle front against the Nazi invaders, has not been forgotten.

Józef Bieniek published a lot of source materials on the subject, including the profiles of more significant Lemko couriers, such as the famous heroism and incredible ingenuity of Stefan Węgrzynnowicz . However, as he admits himself, due to the scattering of the Lemkos, he could not get to much information.

On August 5, 1985, staying with my wife Lucyna and daughter Oliwia in Krynica, in part of the house inherited from my grandfather Harasym Gromosiaku, I recorded a series of conversations with Basil Owl about courier routes through Lemki region and the situation in the area during the Nazi occupation. St. Bazyli Sowa was himself a liaison between the transfer station in Łabowa and Krynica. Until my grandfather Harasym was arrested, he led people to him, to the hideout, which was in the cellar under the barn, from the Kryniczanka stream. In the cellar there was a huge stoneware pot with milk, cream, and beks with sauerkraut. Among them sat Polish officers and other people waiting for a safe passage through the "śtrekę" and the forest to Tylicz. None of the buildings of Harasyma's rich gas survived, but, by chance, the cellar did!

- Bieniek wrote fairly about Lemkos - said Basil Sowa - but he missed a lot because our people left. He could not know about the various outlets, because there was such a conspiracy in us that he knew very little about the other. And nothing was said. It protected us from the stuff. A Pole, sometimes very courageous, drank vodka and talked too much, and on the second day he was hanging behind his back-twisted hands on the ceiling hook at the Gestapo in Muszyna.

Those people from Poland could not come to us, to our outposts overseas, just like ourselves. They were directed from Nawojowa or Nowy Sącz by liaison officers who were especially interested in this. Our people took over and then passed them on. We will never recreate them again, because they deported them to Ukraine, the rest to the west, and the traces of these underground institutions, through which metastatic routes led the way, got lost. These were very complicated activities and always dangerous. At that time, it was not possible to walk along this road. Everywhere there was a German military police, Polish blue police and Ukrainian police - "Sichaczcy" brought here.

Basil Sowa, on my request, gave me information about the institutions he worked with and about people from the Lemko conspiracy he knew. An important role on the metastatic route to Krynica was performed by the Łabowa branch and the institution in Nowa Wieś. in Łabowa, the mayor of Osip (Józef) Wisłocki was in the underground and people from other institutions were brought to him from the side of the Polish villages. In this way they came to Łabowa and Nowa Wieś. In Nowa Wieś, the transfer post was located at Jerzy Steranka . He had a granary from the side of the river, and there people who he was brought to him, he deposited.

- The police sniffed for this because in Łabowa there was already a post of those "sichówków" - told Basil Sowa - and the worst thing was to get through it. There he was a Czech commander. They were not our people. They sent them somewhere from Galicia (that is, from the area of the former Halychyna duchy - SAW), especially from around Lviv. I talked with them, I know.

From Nowa Wieś it was necessary to take these escapees to Kozub, on Werch, because it was a road or something to dream about. A column of Ukrainian police stood at Krzyżówka. Germans and Polish navy police were also standing. They controlled everything, especially milk, which was sent to the dairy for forced deliveries and people often had meat for sale in these convoys. It was hard to get through Krzyżówka through these constant searches. It was necessary to make inquiries over Berest, over the New Village, and so it was necessary to reach this ridge as it ends in the forests behind Berest and then on Mochnaczka, where most were brought to the Greek Catholic priest - Emilian Węgnienowicz, Stefan's father . It was he, from there, that he led these Polish officers to the Chambers, or immediately to Slovakia. There were also other couriers in Mochnaczka and in Izba, who were still running.

I would like to add that Stefan Węgrzynnowicz did nothing to hurt him, that he did not honor him, but how much he gave as a courier. Nobody wanted to remember about it officially, and yet in this service, every day man was exposed very hard, and Węgornowicz was courageous to madness. Life could be lost in heavy pain. I know how it was at Sterankivci, at this Jurek Steranka in Nowa Wieś. Everyone was afraid there, because the whole family could die for maintaining such a point.

It was a beautiful Lemko custom to equip refugees who were crossing the border with food. For ages no one has been let loose on our way without food. It was like this: huge, round country breads were baked, then cut into quarters, and each was cut into a "well". As the bread cooled, butter was loaded into it and covered with crust removed from the cut of crumb. Each of the refugees was given such a portion of bread and butter for the road. This was called "meridia". A nice-sounding word, probably of Romanian origin.

The police, if someone appeared to her suspect, checked if he was true Lemko, by prayer. They then ordered quickly and without a prayer to pray in Lemko or Ukrainian. When someone made a mistake, he stuttered, misinterpreted - it was the end. Setback. This was one of the main reasons why Poles operating in the underground in our area had to be protected by the Lemkos, they had to use their care and instruction.

- It is necessary to correct - he postulated late Bazyli Sowa - what they write after all the years and other publications, that the Lemkos passed Poles across the border to the Slovaks. It's not true. They gave back to the Rusyns there (Lemks). The Slovaks were very uncertain. They were mean people, and the "hliners" were as angry as angry dogs, worse than "sichówków". Our couriers and guides did not trust them at all. We had our own Lemko routes and there were never any mishaps on them. Blunders that occurred in Slovakia to people led by the Lemkos, all took place outside the Slovakian Lemkos.

It remains to supplement and improve information about my grandfather - Harasym Gromosiak. According to oral knowledge, he was in the first three founding ZWZ in Krynica, together with his friend Stanisław Kmietowicz . They were both passionate hunters, they had many personal ties, including once at Harasim's hunt, he saved his life with a good shot when he recklessly ran to the warrior who had been shot. They had boundless trust to each other, which is why Kmietowicz (aka "Storczyk" or "Groszek") drew Gromosiak to a military conspiracy in the area Krynica.

Harasym Gromosiak was arrested together with Sławek Łohazą , his nephew, as a result of denunciation by a provocateur from Muszyna who, being a Gestapo confidant, brought his grandfather a group of Polish officers to transfer. They were kept in prison at Montelupich in Krakow, beaten and tortured. All attempts to extract were ineffective. The family thought that it was true, as the Germans announced, that he had been arrested for hiding a hunting weapon. My grandfather did not release anyone, and Sławko did not know anything, except that he helped Harasym. In the same cell, Stanisław Marusarz, a famous ski jumper from Zakopane, was also caught with them by courier routes. On the eve of the execution, he managed to escape at night, jumping from the second floor. Harasyma Gromosiak and Sławek Łohazę together with other prisoners were shot in the moats of the Austrian fort in Krzesławice near Kraków, now the district of Nowa Huta. The identification of the body was made by my mother, Olga. I was there; the thing took place somewhere in early 1946.

So a new trail! Was the mysterious baca Slavek to be Lemko / Rusin? In the light of what Seweryn Wisłocki tells us, we can be sure that it is. He was a member of one of the smuggling networks operating in Slovakia. Unable to hide Dr. Horak and his companions in Slovakia because of the Germans and Slovak fascists - hliners - he hid them at his finish on the Polish side of the border, carrying them through a transfer channel near Żegiestów or Tylicz. This part of the story has cover in the facts.

There remains one more thing to explain, namely - the meeting of Dr. Horak with Jews - fugitives from the Warsaw Ghetto Uprising (19/4 - 16/1943) and / or (most probably) from the Warsaw Uprising (1VIII - 3 .X.1944).

We remind his diary:

October 30, 1944

We moved slowly, because it was dark on forest paths. For several days we camped in the pine forests and listened to the bang of the department. We saw a group of insurgents who fought with mountain shooters and blue fascist policemen. The fascists withdrew, and we joined the insurgents and we were their guests throughout the day. It was a mixed group from Hechaluts, ŻOB and DROR from the Rzeszów Province in neighboring Poland, who helped our insurgents and could not return - due to deep snow - to their operational area between Krakow and Przemyśl. Their doctor was Rachela W. - widow of a murdered Jewish doctor. She told us about the battles of the group Jesia [Frog] Fryman Bandy with the Nazis and twice fed us with warm food. When these Jewish fighters went north, we had to go south towards Košice, where we arrived on the sixth day. In Košice, we received the order to go to our detachment who was waiting for the Red Army's offensive to join it until the end of the war.

In this context, it may be German Grenzschutz and Polish Blue Police officers. It sounded fantastic, because it was difficult to expect Jews after they were displaced to death camps and liquidated there.

On the Internet, we encountered an interesting article about Jews in the south of Nowosądecczyzna, in which interesting information from our point of view falls:

About the courier route Tarnow - Piwniczna - Slovakia

Jerzy Reuter

In January 1995, next to the municipal cemetery in Piwniczna, three elegant cars parked. It was full winter and visiting the cemetery rather rarely. Three elderly men dressed in black came out of the cars and without looking around, they entered between the graves. They all met at one grave and made a conversation between them. After a few courtesy sentences, they pressed the hats on their heads harder and intoned a sad song in Hebrew, and then they drove together to the house of the Dagnans family .

Piwniczański Dagnans come straight from Stanisław Dagnan, the Tarnów owner of the mill. His son Bolesław moved to Piwniczna after the first war and founded his mill, sawmill and several other enterprises, and uncle Bolesław Jan was at that time a parson priest. Bolesław was a very darting man and in a short time he gave Piwniczna an industrial character, employing in his factories many inhabitants of this beautiful and health resort.

Józef K. Anzelm L. and Mojżesz W. came to Piwniczna to worship the grave of a man who saved their lives several decades ago. That man was Bolesław. Earlier, they took part in the anniversary meeting at the former Auschwitz-Birkenau camp, where they lost their parents, brothers and sisters. History has come full circle and brought its participants to Piwniczna, where many years ago hidden in a mill, they were waiting for an uncertain future. Transported from different corners of Poland, through the stage in Tarnów with Stanisław Dagnan, they waited for their guide and safely crossed Slovakia to Hungary.

The courier route through Tarnów - Piwniczna - Slovakia was established at the end of 1939. Initially, the guides carried out Polish officers heading west, mostly to France, and after several months, the Jewish population was also carried out at the initiative of the Żegota and Home Army organizations. The mills in Tarnów and Piwniczna were perfect hiding places for those waiting for a metastasis. The scale of this dangerous activity is best illustrated by the relation of Michał Łomnicki, one of the most famous guides. Lomnicki, in his memoirs, writes about two hundred Jews carried out by himself and several hundred Poles.

One of the first, or perhaps the first, Jew who was hidden by the Dagnanów family in Tarnów, reached Piwniczna and then to Hungary, he was Anastazy Dagnan's brother-in-law, who was remembered by the guides as Fredek. After paying 200 dollars, he was picked up from the basement mill and brought to Slovakia by the Reichert brothers. Another guide of Fredek, up to Budapest, was Jan Podstawski, one of the most honest Basian couriers.

Adam Bartosz, the director of the Tarnów museum in the article "Zagadka młyna Dagnana" describes the discovery, during the demolition of the building, of a cache built in the attic. The masked room was 9 square meters, and a Jewish family from Tarnów was supposedly hiding there. Did those waiting for a trip to Piwniczna use this box? Or was it still different? We do not know that, but it is certain that the Dagnanów Tarnów mill was an intermediate point on the route to Hungary. This is confirmed by the accounts of the rescued Jews who visit Piwniczna today, as well as the stories of former guides and couriers. The aforementioned Jan Podstawski described his mother Katarzyna, who went to Tarnów after hiding in the Dagnana mill.

One day Podstawska brought from Tarnów a scared young woman and her three-year-old daughter. The woman was the wife of a German official. They took their wedding four years before the outbreak of the war, but after the Nazis entered, they had to immediately divorce and break all contacts with each other. The spouse, wanting to help his family, arranged for them to be transferred to Hungary and, through acquaintance with Stanisław Dagnan, he contacted couriers. After arriving in Piwniczna, the woman and her child lived in the Bolesław mill, waiting to cross the border. Hunger, separation and constant fear for life led the Jewess to a complete loss of self-control and madness. One night, she escaped from the hiding place and boarded with her child on a train going to Tarnow. Maybe she wanted to see her husband again, or she just did not realize the danger anymore. In Bukowa, the SS men entered the train and began to check the documents. As they approached the woman, she took the cyanide out of her pocket and poisoned her daughter and then herself. The ex-husband, at the news of the tragedy, shot himself in his apartment in Tarnów.

Courier routes were many and not only honest people acted on them. There were ordinary bandits, blackmailers and traitors. Perhaps, for this reason, the honest did not try to reveal their stories in the post-war period. Often they were forgotten in their modesty and even moved away from glory. Jan Podstawski, sworn in Budapest together with Stanisław Marusarz by colonel WP Jasiewicz, served in the Home Army under the pseudonym "Rąbalski". He was active under the command of Paweł Libra, pseud. "Clever", the creators of the first metallurgical grid from Piwniczna. After the liberation he lived very modestly and did not impose anyone with his glorious past.

The youngest guide, Michał Łomnicki, arrested in 1942, did not release anyone, despite cruel tortures. Purchased from the hands of the Gestapo, he carried people across the border until the end of the war. He wrote his memoirs and in 1993 he was awarded the Righteous Among the Nations distinction, awarded for saving the Jewish population, and especially for the help given to Henryk Zvi Zimmerman, an activist of the Jewish underground.

The fate of three mysterious men was as devious as could be the fate of people curbed by the perpetrators of the war. Józef K. fell into the hands of the Gestapo in Budapest during the prayer in the synagogue and was transported to Auschwitz. He survived the hell of the camp and lived in Argentina after the liberation. Anzelm L., due to various circumstances, found himself in Yugoslavia and fought in the partisans - he lived in Australia after the war. Mojżesz W. was kept by Hungarian hosts - he lived in Israel after the war.

In 1995, they made their last journey to the past, to Piwniczna and the life they gave them through the Dagnanów family.

For 45 years after the end of the war, the history of Polish Jews and the associated martyrdom were diligently blurred by the then authorities. Many facts have gone away with witnesses of events, many have evaporated from human memory, and the oral transmission from generation to generation is becoming more and more. The Dagnan generation, however, left faith in the size of simple people and the importance of small towns.

/behind the Museum in Piwniczna, Towarzystwo Miłośników Piwnicznej, "Piwniczańscy Żydzi against the background of the city" - W. Wdowiak, A. Talar / An article that appeared in "Gazeta Krakowska / Tygodnik Tarnowski" on 31.07.2009, we publish with

the author's consent. Source: "Polish Righteous" -
http://www.sprawiedliwi.org.pl/?id=123&cid=31/

As can be seen from the above, Dr. Horak could meet with Jewish fighters at the end of 1944 on the Polish-Slovak border. This part of the legend about the Moon Cave also suits historical facts. However, it is small but - we could not find anything about the branch or group of fighters of Jasia Fryman Bandy. It is possible that this is a pseudonym.

Let's summarize. From the above information, it appears that in the vicinity of Żegiestów there were transfer organizations operating in both directions of the border: from Poland / GG to Slovakia and Hungary and vice-versa . Dr Horak's people could pass this way and he himself to Poland, where he was kept in the Moon Cave - a den in a cave at the foot of Wielka Pusta or - which is less likely - in a cave on Zubrzym Wierch.

Why is this location less likely? - it is obvious - three wounded soldiers, including one very hard in critical condition, then the terminal was not able to enter any summit, and the Slavek shepherd, even with the help of his two daughters, would not be able to secretly move them there without paying attention to yourself and other remarks of Germans or Ukrainians. There are no miracles - he could only lead them to forests, and only to a cave that was in the Great Pusta massif. Any other option is very unlikely.

At the beginning of the chapter, we said that there is an interesting and strange connection between the murder of General Władysław Sikorski and the existence of the Moon Cave. Well, we mean the fate of people who are involved in it. Assassins (presumed) by General Sikorski and Cpt. pilot Eduard Prhal , who piloted his Liberator after the war, found employment in the CIA and placed in luxurious conditions in the US or Great Britain.

Captain dr Antonin Horak also went to France and then to the USA, where he ran a bar in the collapsed hole of Arizona or Colorado.

And yet he stood in the official hierarchy much higher than the killers of General Sikorski and Cpt. Prhala! What's the conclusion?

There are two: the first one speaks for the fact that the Americans, who had a specific interest in it, stand above all in Gibraltar. Second application - Cpt. Dr. Horak kept his tongue shut all the time, only when he came to the conclusion that he could do no harm to him - he began to speak. And, paradoxically, this is an argument for its truthfulness. I suspect that even the FBI did not know very well what to do with it - as indicated by the fact that it was allowed to publish his memories. We suspect, however, that they have been deliberately trimmed down and tropes deluged so that no-one would be able to find the Moon's cave in Slovakia on the basis of this account.

CHAPTER XI

Moon Cave - the end of a myth?

So we can say that we know where the Moon Cave is. There are a few indications for this. The first circumstantial - historical:

XX century

At the beginning of the First World War, the Russian army took the Sądecczyzna, but shortly thereafter, they were supplanted with the participation of the Polish Legions. The Polish authorities formed in 1918 covered the area designated by the border of the former Polish-Lithuanian Commonwealth, while Czechoslovakia was formed behind the Carpathians.

The People's Republic of Lemko Republic was also formed, which was trying to prepare the ground for joining the Lemko region to Czechoslovakia. The headquarters of the Lemko movement that was growing a few years earlier was Florynka in the Biała valley.

Just before the Second World War, Polish-Czechoslovak border disputes came to the fore. First, the Poles occupied the Slovak Shovel, the peninsula formed by the bend of Poprad near Żegiestów, and some land near Milik and Andrzejówka, then the Slovaks joined the German invasion. German troops entered Sądecczyzna from the south - from the Red Monastery, Bardejov and (together with the

Slovaks) from the vicinity of Mniszka nad Poprad. [The staff of the Slovak army was in Żegiestów!]

The Polish resistance movement concentrated during the first years of the occupation on the transfer of couriers [from GG to Slovakia and Hungary, and then to London] consciously for their protection, limiting the typical guerrilla warfare. The last year of the war [1944], however, brought numerous fights to the hiding in the mountains of the troops; Julian Zubek, nicknamed "Tatar", became particularly famous, who in 1944 became the commander of the Podhale Rifle Company of the regular People's Polish Army. As a soldier of the Home Army, he was known for his willingness to cooperate with the Soviet partisan group operating also in the Beskid Sądecki. Soviet troops - branches of the 4th Ukrainian Front - occupied Nowy Sącz on January 20, 1945, and the entire Sącz region a few days later.

The end of the war meant the end of the Lemkos in the Sącz region. Most of them were brought to Ukraine by activists who arrived in 1945 with the Soviet army, affecting some of them with promises of well-being, some pressure (some of the displaced people managed to come back). All the rest were displaced in 1947 as part of the "Wisła" action in order to (or under the pretext) to liquidate the backbone of the Ukrainian Insurgent Army operating in the Bieszczady Mountains.

In place of the Lemkos, people from Nowy Sącz, Limanowa and Podhale were brought in, many villages were frozen or shriveled. Returning to Sądecczyzna was very difficult for the displaced until 1980 (it was easier to return to the poviats located further east), so only few people live there today.

Source: Wojciech Nowicki - "Beskid Sądecki - Przewodnik", Pascal, Bielsko-Biała 2001, pp. 17-18.

Second circumstantial evidence:

Marek Kroczek, Tadeusz Wieczorek
Natural route after matresses of the Karpacka District name Adam Count Stadnicki

One of the forms of area protection of nature and the wider landscape in Poland are Landscape Parks. In the Beskids, two such parks have been created so far - Żywiecki Landscape Park and Poprad Landscape Park. The Poprad Landscape Park was established on September 11, 1987. It covers an area of 54 thousand. ha, and with the buffer zone 78 thousand. ha, being one of the largest landscape parks in Poland. Almost the entirety covers the area of Beskid Sądecki. The aim of creating the Park is comprehensive protection of natural, landscape, spa and tourist values realized by adapting business activities to the requirements of nature conservation. The creation of this Park in the area developed, but with high landscape values was to be an example of the co-existence of forest and tourist economy based on the rational use of existing natural resources.

Among the natural factors, the leading feature in his landscape is the relief of the land conditioned by the geological structure. Morphometric features of the sculpture (slopes slope, relative and absolute heights) condition the availability of the land, which directly or indirectly affects the behavior of natural landscapes, or similar to them, and the shaping of the cultural landscape. Physiographic reasons of the Poprad Landscape Park and historically conditioned settlement determined the manner of land use. Almost 70% of its area is occupied by forests and only 27% by arable land.

As a result of work on the creation of a pan-European integrated system of protection of the natural heritage of the continent as a whole, ie the European Ecological Network EECONET (European Ecological Network), the area of Poprad Landscape Park has gained the highest rank in the valorisation of protected areas.

It is a nodal area of international importance in the National Ecological Network with biocentries and buffer zones. The Dunajec River has been recognized as an ecological corridor of international importance.

There are as many as 12 forest reserves on 14 nature reserves established so far in the Sądecki Beskids. Some of them have their beginnings in the rational activity of the former owner of most of the forests of this region of Count Adam Stadnicki.

The oldest nature reserve was created in its estates around 1906, and the youngest in the State Forests in 1996. Some of the reserves are easily accessible (Łabowiec, Barnowiec) or even made available (Las Lipowy Obrożyska), but there are also reservations difficult to penetrate and not very often visited (Hajnik, Żebracze).

The current area of reserves is about 500 ha, supplemented with almost 90 natural monuments.

The aim of the nature trail is to familiarize young people and tourists visiting this beautiful part of the Poprad Landscape Park with the values of magnificent and unique mountain nature, impressive valleys of landscapes with clear streams of water, the charms of sunny logs stretching among forests. It is also the answer to the large tourist penetration associated with the opening of the gondola lift to Jaworzyna Krynicka.

Marked on the route of the stops allow you to thoroughly familiarize yourself with the stipulated diversity of vegetation, with the world of animals, indicate the most valuable forms of terrain and interesting hydrological phenomena. The beginning of the nature trail is located in the center of the village of Łabowa, next to the PKS bus stop on the Nowy Sącz - Krynica road. The name of the town is supposedly derived from the Elbe - the Russian nobility, who settled here after the Tatar invasion. Łabowa was famous for boisterous fairs that were taken from all over the area. Formerly, the village was inhabited by Lemkos. The trail runs from the bus stop to the west. We cross the bridge on the Nawojowska Kamienica river. At the bottom of the river and along its shores, the Marseilles are revealed. Approximately 200 meters long at the bottom and the sandstone complex at the edges, slate-thin slabs distribute marl deposits. Sandstones are finely and medium grained, poorly sorted with a limestone bond. They contain, in addition to quartz, also muscovite, glauconite and quartzite, and less frequently crumbs of limestone and feldspar. On the right we pass a wooden church from 1930 with a historic painting of Saint. John of Nepomuk from the late seventeenth century and late baroque statues of St. Luke and Saint. Mark.

We reach the intersection and turn left onto the asphalt road. After about 500 meters on the left we can see a beautiful brick church from 1784 with a tower with an onion mask, funded by the Lubomirski family. After about 10 minutes, we reach the Uhryński Potok valley. On the left, there are beautifully forested slopes of Ośnikowski Wierch (828 m above sea level), on the right side of the Peresliha slope (812 m above sea level).

We can see here a deeply indented (sections up to 30 meters) winding valley Uhryńskiego Potok, overgrown with mixed forest with handsome pine specimens (designed geological reserve). This area is best accessible from the south-east, by a forest path that leaves the trail at a distance of about 100 meters below the cave house. This path runs over the bends of the stream and after its second crossing it leads to a valley with relatively gentle slopes. From here, start hiking up the stream with its meandering trough, which is possible only when the water level is low. At the bottom of the stream and its steep shores, and especially in the eroded bends, the profile of the Lower Eocene is almost continuously exposed. These are thin - or medium-sized sandstones with a calcitic arrow translated by shale. Further up the stream reveals a complex of clay shales with a few thin schools of sandstones. In the bottom part of the profile there are free-falling slates, usually red, which contain numerous holes. In the upper part of the profile, the characteristic work is the so-called striped uniform which consists of alternating thin layers of chocolate - cherry shales and marly clays with a blue - gray color. The latter dominate in the highest part of the complex. The wooded Uhryński Potok valley has extraordinary landscape values, which include its significant depths, winding course and steep banks, as well as almost vertically deposited multicolored rocks exposed at the bottom of the stream and on its banks.

After passing about 4 kilometers from the beginning of the route, on the right side there is a camping place prepared by the Nawojowa forest inspectorate. You can rest here on the benches, and also look for mineral water springs in the nearby stream, which beat at the bottom of the river.

Going further after about 2 kilometers, we reach the village of Uhryń. There are interesting half-timbered chapels in it. At the beginning of the village, we can see a renovated chapel with a very interesting construction. We go through the whole village, about 500 meters behind the last buildings there is a barrier. From this place, we can see the entire route from Łabowa.

We go further along the asphalt road. After about 2 kilometers, we reach the road fork, turn left to the Uhryń reserve. On the left side of the path we can observe an interesting fir tree. From two trunks after joining at a height of about one meter, one joint stem is formed. We come to the wood warehouse. About 100 meters behind it, we reach the border of the Uhryń reserve.

Uhryń Reserve - was established in 1924 in the private property of Adam Count Stadnicki , confirmed by the decree of the Minister of Forestry and Wood Industry from 1957. The former name of the reserve is Medwiczka. The area of the reserve is 16.00 ha. The protection covers beech and fir stands, about 230 years old, characteristic of the old Carpathian forest. In the stands firm with a small share of beech and smaller amount of spruce. The oldest firs reach a height of up to 40 meters and trunk thicknesses up to 115 centimeters. In the swamp most often there is black-necked blackberry, female louse, gamekeeper and gamekeeper. From protected plants, there is a hive in the hive, and partly protected - a common hoarfrost and sweet woodruff. It is one of the oldest Polish reserves and for over 100 years there was no economic activity in it. The reserve is located within the Nawojowa Forest District.

What follows from the above? And the fact that in the Beskid Sądecki there are geological formations described by Dr. Horák: sandstones lying on carbonate rocks!

And next:

We continue along the path along the border of the reserve (leave the reserve on the right). After about 15 minutes, we come to the celebration path. Above the rock outcrop where we can rest for a moment and look deep into this Primeval forest complex. Above we can see a young tree stand, typical succession in the Carpathian forests. After passing the next 10 minutes, we reach the ridge and bounce left along the border of the reserve. We reach the intersection.

There is a place to rest and shelter in case of rain. On the right, leave the reserve, head straight up the slope and after 200 meters turn left after the arrow. After about 20 minutes we reach the red trail. The trail turns right in the direction of Hala Łabowska, we are located about 4 kilometers from it. After about 40 minutes, we reach Hala Jawor, which falls to the south-east in the abysmal junction of the Potaśni tributary. The hall is now overgrown with a beautiful pine forest. We go further north, after a while we come to the monument commemorating the death of the Home Army partisans, then to Hala Łabowska and to the right to the Łabowiec Reserve.

Hala Łabowska (1061 m above sea level) is a vast overgrowing forest with a dorsal hall descending towards the north, located near the keystone, where the side ridge of Parchowatka (1005 m above sea level) joins the main ridge from the south. The view from the hall is indeed limited from the north and north-east, but very varied.

Ahead of us, behind the Kamienica Valley, there is the range of Margoni Niżna and Wyżna, Spalska Góra, Tokarnia and Czerszla, with the Jaworze and Chełm banks sticking out of it, to the right of it. The hall was closed after the last war - thanks to the efforts of the PTTK Branch in Nowy Sącz - a hostel, which on the Main Beskid Trail, between Krynica and Rytro constitutes an important point of support for tourists. In June 1944, during the Gestapo raid, Adam Kondolewicz , a soldier of the "Tatara" Home Army, died in the hall , where he died, a boulder with a commemorative inscription: In the hostel we can eat a meal (very good kitchen), and spend the night before further walking along the nature trail The Hala Pisana is the name of the gable hall that currently covers the main ridge between Zadnim Góry on the west and Wierchem nad Kamieniu on the east, this name comes from the time when geodecs made measurements in the eye and "wrote down" how the highlanders said. There is a commemorative boulder in the hall reminiscent of the Nazi occupation and the struggle of partisans who have their hiding places in the forests of the Beskid Sądecki. Above the beech forests that surround the hall - the original view of the west on the nest of Radziejowa and Prehyba, distant Gorce and Beskid Wyspowy and Babia Góra in the most distant plan. Towards the north, a plastic map of the Sądecki Basin appears, and towards the south-east, beautiful Parchowatka, the surroundings of Jaworzyna Krynicka and Runka. To the south-west, with gentle weather, gently drawn Tatra Mountains.

Thus, the area was active in terms of the activities of partisan groups during the Second World War. Traces of this are commemorated to this day!

We continue along the red trail, accompanied by yellow signs. The signs lead to a dark, damp forest in which old, ancient beeches prevail, full of darts and boulders covered with vegetation. When the north-westward path breaks and turns right to the west - near, to the left, a bit to the bottom - the spring. Soon, we leave the forest to the hall, which culminates Zadnie Góry (969 m above sea level), remain slightly on the left. A panorama from it is extensive and it is worth going to the left on the southern side of the ridge to view the Poprad Valley and the shaped Parchovice. There is a ridge ditch, rare in the Carpathians, so-called a double ridge, an example of the processes of dismembering mountain ridges. It is accompanied by numerous sculptures typical of landslides, such as rock walls and pulpits, clefts, mobile blocks, dilatation gaps and colluvial swellings.

From Zadni Góry we head west to the junction of trails, then turn right and follow the green trail to the northwest. After about 10 minutes, we reach the blue trail that leads us towards Makowica. Under the peak, the nature trail departs from the tourist route and descends to the western slopes of Makowica, on beautiful glades. There is a lookout point here, from where we can see a beautiful panorama of the Radziejowa Range and the Poprad Valley. From Lego, we follow the glades and the forest into the valley of the Rzyczanowski Stream. Coming up to it, we turn left and follow the forest path that runs along the bottom of the valley.

We reach a beautiful wooden bridge over the picturesque gorge of the Rzyczanowski Stream. It flows on this section with a rock gorge of a depth from a few to a dozen or so meters and a width in the bottom part of 5-10 meters, and in the upper part from a dozen to 50 meters. This gorge was eroded in the thick-bedded and very thick-bedded

Magura sandstones. The existence of fragments of several levels of valley flattening, including terraces built of river sediments, testifies to the multi-stage development of the valley. The cutting of resistant sandstones and the creation of ravines is a testimony to strong erosion in the last Holocene stage of its development. The winding, meandering course of the modern channel is perhaps inherited from the older stage of valley formation. Magura sandstones create attractive landscapes and morphologically interesting cliffs and rock performances in the walls of the gorge, while its bottoms are small waterfalls, cascades (rock stages) and shypot (small protrusions on outcropping banks separating the river bed). The most attractive landscapes include: the lower rock gate, the rock gate at the bridge and the top gate. The gorge of the Rzyczanowski Stream is also the site of the contemporary formation of limestone necrosis.

The most interesting of the deadwood covers is formed on the steep slope of the ravine, below the bridge. Perhaps its creation is associated with the outflow of waters enriched with CO_2 and strongly mineralized, migrating along the fault zone. The best viewing point is a wooden bridge over a stream at a height of about 25 meters. We can see the charms of inanimate nature, but also a beautiful beech-fir tree stands along the stream. He is over 80 years old. You can find many protected plants in it. The entire area is a surface nature monument. Behind the bridge there is a place to rest.

Again, the convergence with the descriptions provided by Dr. Horák.

The rest of the route runs along a forest road towards the village of Rzyczanów. At the first buildings, turn left, going down the tourist trail. We go through the entire village, heading towards Poprad. At the end of the village there is a beautiful, renovated chapel. We reach the bridge in Poprad. After the bridge we turn left and along the Stary Sącz road - Piwniczna we go to Rytro. It is a summer village located at the mouth of Roztoki to Poprad. On the steep, conical hill on the right bank of the river, the ruins of a knight's castle. The name Rytro is already mentioned in 1240. In the thirteenth century, the castle was owned by the castellan of Nowy Sącz, Piotr Wyżga, who sold himself to the Teutonic Knights and, as a punishment after his death, he supposedly suffered in the ruins of the castle.

(Incidentally, Piotr Wyżga was also an alchemist and gold seeker, hence an extremely interesting figure...)

Here we say goodbye to the nature trail Adam Count Stadnicki after the forestry of the Carpathian wilderness. We cordially invite you to visit the trail and get to know the beauty of the Poprad Landscape Park.

Source: www.nsi.pl/almanach/artmiejsca/szlak_przyrodniczy_hrabiego_stadn ickiego.htm

* * *

And do you know what's going on? And it follows that we have mistakenly interpreted Dr. Horak's data! Well, he and his men jumped back to the Germans and hliners, and then they went north - towards the border with GG - that is, with Poland. The towns that he named were the last known on the Slovak side of the border - that's why he remembered them, because he went towards Sulin or Lipnik, where they crossed the border in Poprad. Then Slavek helped them - actually a Lemko guide with their daughters who lived in Yzdar - Izdworze near Wierchomla Wielka. He could not stop them because it was probably a smuggling meta (den) so he took them to one of the guerrillas hiding places - near the Łabowa Hall. They were empty, because the partisans in the meantime went to the East and joined the Soviet Army - this still needs to be checked, but I think that in October and November 1944 there was no one there - not even the

Robert K. Leśniakiewicz, Miloš Jesensky

shepherds. Slavek led them to the known mark - Bear's Cave. And indeed - it is known that the bones of a cave bear were discovered there. In a straight line, it is only 6 km from Yzdar, so Slavek and his daughters could visit them in this cave.

The escape route of Dr. Horak and his two companions leads the most logical route - on mountain ridges and forest tracks. Today they have grown thick, but in the 1970s they were still used for logging or wood transport.

What happened to Slavek? Most probably, he went to Slovakia in 1945-47 and the trail left him. He could work for the Home Army, the British and the Soviet partisans. Such people were worth their weight in gold. He could also work for Slovak patriots fighting in the SNP.

Also, the geological structure would correspond roughly to the relation of Dr. Horak - there are also sandstones of carbonate rocks outside the flysch sandstones.

In 2010, we finally started searching for the Moon Cave. We made a direct assumption that it is one of the larger caves of Beskid Sądecki. Thanks to the help of the President of the Jordan Land Lover Society, mgr inż. Stanisław Bednarz obtained a list of caves of Beskid Sądecki, from which we selected the following formations:

LIST OF THE BIGGEST FAVORS OF BESKID SĄDECKI

(with length of corridors> 10 m)

Inventory number Name Length Denialication

K.Bsd-01.01 Siekier's Cave 26.0 -8.0

K.Bsd-01.03 Dziurawa Well I 11.0 -3.0

K.Bsd-01.05 Wietrzna Dziura 43.0 -12.0

K.Bsd-01.07. Concluded Well 10.0 -8.0

K.Bsd-01.12 Roztoczanska Cave 140.0 -10.0

K.Bsd-01.13 Cave Where the Tanks fell 17.0 -5.0

K.Bsd-01.14. Angry Slit 11.0 -4.0

K.Bsd-01.17 Bania in Radziejowa 24.0 -9.0

K.Bsd-01.18 Jaworznicki Bell 20.0 -11.0

K.Bsd-02.07 Hole after Terminal 16.0 -4.0

K.Bsd-02.08 Szczelina in Urwisko 20.0

K.Bsd-02.14 Jaskinia Złotniańska 155,0 - 12.0

K.Bsd-02.20 Stone Sculpture Cave 12.0 -7.0

K.Bsd-02.22 The cave in Cracked Copy 13.0 -5.0

K.Bsd-02.23 The cave of St. Step 41.1 -11.0

K.Bsd-02.28 Bear's Cave 340.0 -28.0

K.Bsd-02.31 Feleczyńska Hole 46.0 -15.0

K.Bsd-02.36 Grota nad Pusta Wielki 3.0

K.Bsd-02.37 Smerczynowa Hole 17.0 -3.0

K.Bsd-02.41 Czarnopotocka cave 20.0 -7.0

K.Bsd-02.38N Zbójnicka Cave 57.0 -7.0

Only three of them, which were distinguished with bold letters on the list, fit in with everything written about the Moon Cave. The length of their corridors is over 100 meters and their denivelation is relatively large - it ranges from 10 to 28 meters.

Some of them, such as Smerczynowa Dziura in Zubrzym Wierch were known to the local population. It is located in the saddle between two peaks of Żubr Wierch. A horizontally situated cave opening is located in the middle part of the saddle, on the northern edge of the overgrown forest clearing. Inside the cave we get through a meter deep sump and a tight stile at its bottom. The first 5 m is a high corridor with a decreasing bottom. Through a low transition in the ruby you can get to the next party. After 8 m, the wide and high (on 2 m) corridor closes the rubble again. (This is where the passage to the further part of the cave could have been up to the Crescent-shaped

Shaft). Beech leaves are located here, which have entered through a narrow slot extending to the surface in a shallow funnel. The cave was formed in the thick-layered Krynica sandstones (Magura series). The bottom covers the clay in the rum. The light reaches into the immediate vicinity of the hole. Fauna and flora were not observed.

Smerczynowa Dziura was first described by A. Rotter in 1976. At least from the beginning of the 1980s, the opening and the entrance well were covered with dirt, and the cave was inaccessible. According to the residents of Wierchomla, "she made a deal". On January 21, 1990, Smerczynowa Dziura was unblocked and scarred by E. Berk and K. Faron . In October 1994, E. Berek defeated the final flood, after which he discovered a 4 m long corridor, ending with a fissure running towards the surface[31]. This is the first "suspicious" cave in which Dr. Horák and his companions could stop. The second is Grota nad Pusta Wielka.

The word "grotto" adequately reflects what this hole really is. It is located at an altitude of 1050 m above sea level, in the massif of the Wielka Pusta mountain. You can reach it from the summit of Wielka Pusta - a ridge about 50 m towards the west, to the foot of the pulpit. The slotted hole No. 1 is located in the northern base of the rock.

The hostel is formed by a high and narrow gap leading from the hole No. 1 towards the south. After 2 m it changes direction at a right angle and through 1.5 m a high peak leads out through hole No. 2 to the surface. The hostel was created in the thick-walled Krynica sandstones of the Magura series. The light reaches to the end. A viper was found in the cavity opening. Other animals were not seen. A hostel known to the local population and was described by speleologists.[32]

---

[31] M. Pulina - "Caves of the Polish Flysch Carpathians", Warsaw 1992.

[32] Ibidem.

Assuming that Dr. Horák and his people could not go far to the mountains and make strenuous marches due to wounds, it can be presumed that they could only reach one of these two caves and find shelter in them. Perhaps it is at the Smerczynne Hole that there is a further level - or even several levels forming the entire cave system - in which was also the Moon Shaft.

Unfortunately, we have no way to investigate.

There is also another possibility that Dr. Horák and his men were moved across the border, not in the area of Żegiestów, but in the area of Mount Eljaszówka, and led to the Roztoczańska Cave, under which one should seek the formation of the Lunar Shaft.

Other possible locations are the Złotniańska Cave and the Niedźwiedzia Cave, which during the war until the summer of 1944 were the bases of the AK partisans. Slavek led them to an abandoned base in one of these caves, and it was there that Dr. Horák made these amazing discoveries. And then the cave "has formed" - already due to the earthquake, it is already due to the detonation of TNT loads by people who know about their existence, so that they do not fall into the hands of Poles or Russians.

The area was penetrated by speleologists and naturalists in the 70s and 80s of the last century and something as big as the cave system would not have escaped their attention. Therefore, one can safely assume that if the Moon Cave really exists, it can only be found in the Polish Beskid Sądecki or in the Slovakian Lubovnańska Vrhovina. There are no other options.

ABOUT AUTHORS

| ROBERT K. LEŚNIAKIEWICZ | Dr. MILOŠ JESENSKÝ |

Robert Konstanty Leśniakiewicz (born on June 7, 1956 in Szamotuły) - Polish writer, translator and publicist, ufologist and researcher of other phenomena from the so-called "Borderline knowledge", also deals with environmental protection. A graduate of the Officer's Mechanized Forces Academy in Wroclaw. Reserve captain of the Forces of the Border Protection (1975–1991) and the Border Guard (1991–1994), civil specialty - forestry, forest protection. He currently lives in Jordanów.

From 1987, a member of the Space Contact Club and the Center for Research on Anomalous Phenomena in Krakow and a correspondent member of the Brazilian Centro Brasiliero de Pesquisas de Discos Voladores (since 2007), as well as the vice president of the Jordan Land Lovers Society in Jordanów, a member of the Mushroom Lovers Club, founding member and member (until 2010) of the board of the Mushroom Lovers Association in Krakow.

As part of his literary activity, he collaborated with the monthly magazines: Granica (1987–1991), Sfinks (1989–1991), the UFO quarterly (1990–1998), the monthly magazines Eko Świat (since 1994) and Nieznany Świat (since 1991) - and co-edited (together with Bronisław Rzepecki) the quarterly Peripheral Visions, UFO Time and World UFO (1997–2000), for which he wrote over 2000 articles and translations from English, Russian, Czech, Slovak and Swedish. He also co-edits the regional quarterly "Echo Jordanowa". He collaborated with Andrzej Zalewski as part of the Ekoradio broadcast of Program I of Polish Radio. He is the author and co-author of 31 books and translations from five languages, and runs two blogs devoted to UFOs and paleo-astronautics, ecology and other paranormal phenomena.

Dr. Miloš Jesenský Ph.D. (born on May 23, 1973 in Čadca) is a Slovak researcher of anomalous phenomena, ufologist, writer and publicist. He studied at the University of Veterinary Medicine in Košice until 1995, then until 1999 he worked as an independent employee of the History Department at the East Slovakia Museum in Košice. From 1998, he was an independent researcher at the Faculty of History of the Slovak Academy of Sciences (SAV) in Bratislava. His doctoral thesis was entitled History of Alchemy in Slovakia, in the years 2001–2006 he worked in the Žilina Library as a bibliographic specialist, and then in the Office of Information and Foreign Contacts in the local government of the Žilina Region. Currently, he is the director of the Kysuce Museum in Čadca.

Dr Jesenský specializes in journalism in the field of history riddles, both in Slovakia and abroad. His literary output is extremely rich, he has written over 40 books, and his articles are published on websites and in magazines, also in Poland. He is a member of the Association of Slovak Writers and the Slovak Pen Club as well as the Syndicate of Slovak Journalists. He is also the winner of the Crystal Tiger Award (1998).